PRINCIPLES OF CLIMATOLOGY

A MANUAL IN EARTH SCIENCE

HANS NEUBERGER
JOHN CAHIR
The Pennsylvania State University

HOLT, RINEHART AND WINSTON, INC.
New York Chicago San Francisco
Atlanta Dallas Montreal
Toronto London Sydney

Library of Congress Catalog Card Number: 75-85751
SBN: 03-078800-5
Printed in the United States of America
1 2 3 4 5 6 7 8 9

PRINCIPLES OF CLIMATOLOGY

A MANUAL IN EARTH SCIENCE

TO OUR PATIENT WIVES

preface

This introductory book was written to close a gap in existing text material used for the earth science courses that are ordinarily offered in schools of education, state colleges, liberal arts colleges, and community colleges. The study of climatology provides valuable background for understanding the ecology of our environment, geographic and geologic phenomena, hydrologic events, and the behavior of oceans. In short, climatology is an important requisite for the other earth sciences. Furthermore, the subject is of interest in its own right, as its long history attests.

Climatology has evolved from a purely descriptive adjunct of geography into a physical science. With this development, the methods used in climatological research have changed from the most primitive statistics to highly sophisticated mathematics, from the static to the dynamic approach. Thus, climate is no longer considered an invariant manifestation of the atmosphere in contrast to the variable weather, but a changing composite of functional relationships, each susceptible to separate analysis, but none entirely independent of the other. Of course, the time scale of climatic fluctuations is considerably larger than that of meteorological ones. Nevertheless, climatology is closely related to meteorology, and real understanding of it involves meteorological concepts. For this reason, we have emphasized the principles and controls of climate, that is, the meteorological ones, leaving to the classroom instructor and the students the discovery of the climatic facts as they relate to individual places.

It might be remarked that the exercises interspersed throughout the text are an integral part of it. For some of the exercises, prepared graphs are provided to eliminate "busy work" that would not add to learning. Yet, we felt that the graphing of numerical values will induce the student to pay closer attention to trends and magnitudes. Only after completion of the exercises does the text form a closed unit. This innovation in

presenting climatological material was made in order to involve the students more directly in the teaching–learning situation.

Certain facts of climatology such as average global distributions of pressure, temperature, and so on, are not stressed in this book. For this reason, it might well be supplemented by readings from a good regional geography or climatology text.[1] We believe that the facts alone do not convey to the students an adequate picture of the different climates found on the earth. For the same reason, climate classifications were omitted; in this connection, it is perhaps worth recalling that Köppen, the originator of the most widely known classification system, once remarked that "a classification is necessary for the understanding and clarification of a complex subject matter, but once the subject is understood, the classification becomes superfluous."

The appendixes are intended as sources of information on: 1) some samples of simple experiments that can be used as pertinent classroom activities at the appropriate point in the discussion; 2) directions for the construction of inexpensive instruments and methods for making climatological observations; 3) cookbook-type statistical schemes for meaningful evaluation of climatological records; 4) a review in the form of questions and problems using comparative climatic data of twenty stations in the conterminous United States. The information contained in the appendixes can also be used for planning science projects in climatology.

The material presented here has been tested three times as a text for the course "Physical Climatology For Teachers" offered at the College of Earth and Mineral Sciences of The Pennsylvania State University as part of a sequence of teacher courses in earth science. We have been encouraged and guided by the students in this course, most of whom were teachers themselves, and we wish to acknowledge their contributions.

We are indebted to the Literary Executor of the late Sir Ronald A. Fisher, F.R.S., to Dr. Frank Yates, F.R.S., and to Oliver & Boyd Ltd., Edinburgh, for permission to reprint Table III.4 from their book *Statistical Tables for Biological, Agricultural and Medical Research.*

Hans Neuberger
John Cahir

[1]See list of suggested reading on p. 135.

EDITOR'S NOTE

The reader will find a number of blank pages on the reverse sides of pages with exercise diagrams or tables which are to be completed. This arrangement enables him to answer the exercise questions on the blank spaces, detach these pages for correction by the instructor, and replace them afterwards in proper sequence, using a ringbook as binder. For exercises that require no drawing of diagrams the student will furnish his own paper of suitable size for insertion in the text.

contents

PRINCIPLES OF CLIMATOLOGY

A MANUAL IN EARTH SCIENCE

1 introduction

In centuries past, few people traveled over great distances and experienced climates substantially different from what they were accustomed to. Furthermore, because transportation was slow, the transition from one climate to another was very gradual and, therefore, the stress of acclimatization was minimal. In the modern air and space age, mobility is accepted in many populations; this, combined with near sonic, and soon supersonic, speeds has increased the awareness of the climatic variation observable on our planet. A business executive may leave Kennedy International Airport on a chilly spring day and land in 12 hours at Bombay, India, where the temperature may be higher by 50°F or more with the relative humidity near 100 percent. Climatic shocks experienced when stepping out of an air-conditioned aircraft into the humid heat at the onset of the summer monsoon may be literally breathtaking.

The growing industrialization of developing nations in widely separated regions has fostered commercial exchanges on a scale that was unthinkable less than 100 years ago. Recent wars have also forced untold numbers of people to get acquainted with, and adjust to, new and often hostile climates. While climate has always been an important factor in the total ecology of a region, especially as a determinant in the agricultural pursuits of earlier times, the requirements of a technological age widen the scope of climatological relevance very significantly.

In a strict sense, climatic variations are found over very short distances; New York City's climate is different from Newark's. However, substantial insight into the more drastic variations discussed above can be gained by a consideration of the large-scale controls on climatic regimes exerted by the geometrical relationship between earth and sun, the nature of the surface, the topography, the availability of water, and certain physical constraints on a rotating planet under unequal heating. In the following pages, climatic variation will be examined from the point of view of the large-scale controls; local variations will be viewed as second-order effects superimposed upon the primary regime prescribed by those controls.

The existence and importance of large-scale controls imply that human control of climate is far beyond present technology. Nevertheless, man has inadvertently achieved certain local climate modifications, for example, by damming water to form substantial lakes, by altering the landscape, by changing the composition of the air with pollutants, by changing the air temperatures around heat sources in urban and industrial areas. These modifications may pale in comparison with future ones. According to the National Academy of Sciences, it is possible under certain circumstances to increase rainfall significantly. Modest beginnings have been made on other aspects of weather modification, one of the most notable being the dissipation of some types of clouds and fog over airports. Still, the large-scale natural controls will be operating.

To attempt a definition of climate seems both futile and unnecessary;

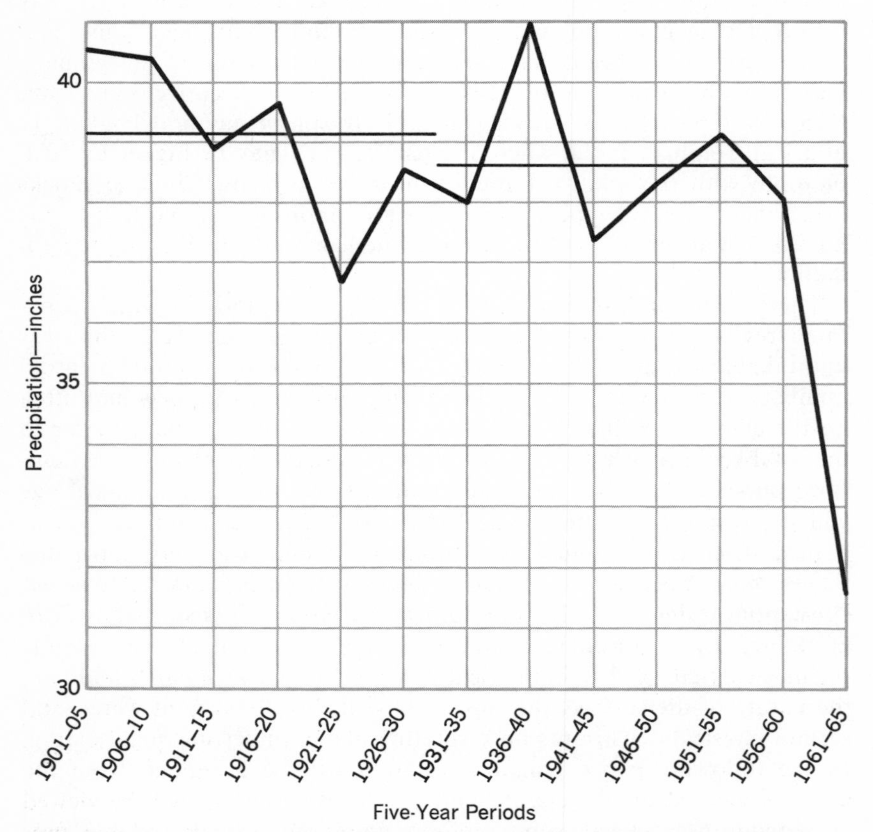

FIGURE 1.1 Five-year period means of annual precipitation at University Park, Pennsylvania. (The standard normals for 1901–1930 and 1931–1960 are shown as straight lines covering the respective periods.)

futile because no single definition could include all the possible ramifications, unnecessary because everyone has a satisfactory notion of what climate is. To say that "climate is the average weather experienced at a given place during a specified period of time" merely shifts the task to the impossible definition of "average weather." Further, averages do not adequately represent any climate; Portland, Maine, and Portland, Oregon, have similar annual average temperatures, but their temperature "climates" are very different.

In any case, it is necessary to make some specifications of the time scale when discussing mean values. Long-term changes in climate do occur, requiring cautious comparisons among sets of stations having records of different lengths. This problem is avoided by calculating means for all stations over the same recent period of sufficient length to obtain reliable values. According to the World Meteorological Organization, such means are called "period means" for which the years used are to be named. The "normal" is the mean value computed from any 30 consecutive years. The currently applicable "standard normal" is the 30-year mean computed for the period 1931–1960. The previous standard normal was for the period 1901–1930, and the next one will be for 1961–1990.

In Figure 1.1 successive five-year period means of annual precipitation at University Park, Pennsylvania, are reproduced together with the two pertinent standard normals. It is interesting to note that the standard normal has decreased during the middle third of the century; the recent persistent drought, unless balanced by an equal excess of rainfall during the next two decades, will continue this downward trend in the standard normal.

EXERCISE 1

Variation of the standard normal affects the interpretation of the departure of individual values from the standard normal. For example, the annual precipitation for 1915 and 1939 at University Park was 38.67 inches. The standard normal for 1901–1930 was 39.17 inches; for 1931–1960 it was 38.63 inches.

a) Compute the departures of rainfall for these two years from their respective standard normals, positive departures referring to the cases in which the observed rainfall is larger than the respective normal.

b) What are the departures from the respective five-year period means, the one for 1911–1915 being 38.92 inches, the one for 1936–1940 being 40.88 inches?

There is evidence that the northern hemisphere is undergoing a slight cooling trend since the middle of this century, following a warming trend of the order of one or two degrees Celsius (centigrade) for the first half of the century. On a much longer scale, the present geologic era exhibits temperatures somewhat higher than the average of some 10,000 years ago. Unfortunately, there is no possibility of studying long-term trends from direct observations, because continuous records of elements such as temperature and precipitation for periods of 200 years or so are available for only a handful of locations. Widespread observations for the inhabited regions of the earth have been made for only a few decades and are still lacking in some of the developing countries.

A further consideration of significance in any discussion of climate is the question of areal scale. The climate is really different at every point, but climatological comparisons can be made over a continuum of distance scales ranging from a few inches to distances between continents. Indeed, the climate of Earth can be compared with that of other planets. Some authors make arbitrary scale distinctions such as microclimate, mesoclimate, and macroclimate. The point of view taken in this discussion is that the scale must be specified; the emphasis remains, however, on large-scale controls.

In terms of vertical scales, there is, and should be, a great preoccupation with the layer of air a few meters deep next to the ground. One of the interesting conclusions that can be quickly drawn from a study of this layer is that its climate is very heavily influenced by the character of the underlying surface, more so, in fact, than by the rest of the atmosphere above. But this is not to say that climatology reduces to a geographical study of various surfaces. It is just the more subtle, or second-order, interactions with the rest of the atmosphere that go far to determine the actual climate; the crucial point is that the underlying surface prescribes the range of possible distributions of climatic elements.

2 the primary control of climate

2.1 GEOMETRIC RELATIONSHIPS BETWEEN SUN, EARTH, AND ATMOSPHERE

The sun is the basic energy source not only for all life on earth, but also for the physical processes in the atmosphere which in their totality determine the climates. For our purposes, this source can be considered constant, because the changes in radiant energy output from the sun, as manifest in sun spots, solar flares, and in similar phenomena, are relatively small and short-lived, at least in the present era. Changes of considerably larger magnitude may have occurred in past geological ages and may have been responsible, at least in part, for the large climatic fluctuations, such as the ice ages, which have been deduced from geologic records.

The average energy received by one square centimeter of surface exposed perpendicularly to the sun's rays outside the atmosphere, when the earth is at its mean distance from the sun, is two calories per minute. This quantity is called the "solar constant" (approximately 2 cal cm^{-2} min^{-1} or 2 langleys per minute, abbreviated ly min^{-1}) and is an amount that, by definition, would raise the temperature of one cubic centimeter (= one gram) of water by 2°C in one minute. Transport of a quantity across a unit area in unit time is called the flux of that quantity; thus the solar constant is an energy flux.[1]

In defining the solar constant, the specification of the mean distance of the earth from the sun is required by the earth's elliptical orbit, because the energy flux from any source is inversely proportional to the square of the distance from that source. It is therefore relevant to inquire into the extent to which the varying distance between the earth and the sun

[1]Flux is sometimes taken to mean the quantity transported across any surface per unit time.

changes the amount of energy flux available to the earth in the course of a year.

EXERCISE 2

Let us denote the solar constant by $D_0 = 2$ ly min^{-1}, the earth's mean distance by $S_0 = 149.5 \times 10^6$ km. According to the "inverse square law," the energy flux S_1 received when the earth–sun distance is D_1 can be determined from

$$\frac{S_1}{S_0} = \frac{D_0^2}{D_1^2} = \left(\frac{D_0}{D_1}\right)^2$$

or

$$S_1 = S_0 \left(\frac{D_0}{D_1}\right)^2.$$

a) What is the energy flux on 5 July when the earth is at the greatest distance of 154.5×10^6 km?

b) What is it on 3 January when the earth is at the smallest distance of 144.5×10^6 km?

c) What are the respective percent differences from the mean? (This is found by dividing the departures from the mean by the mean itself.)

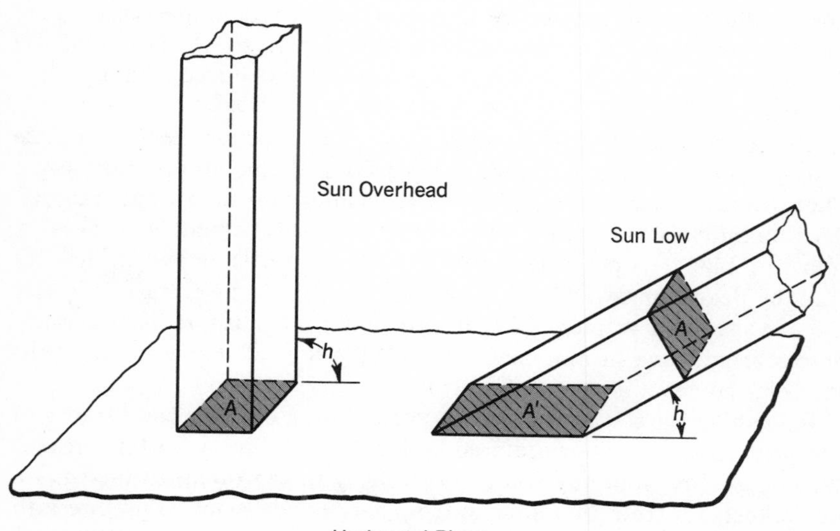

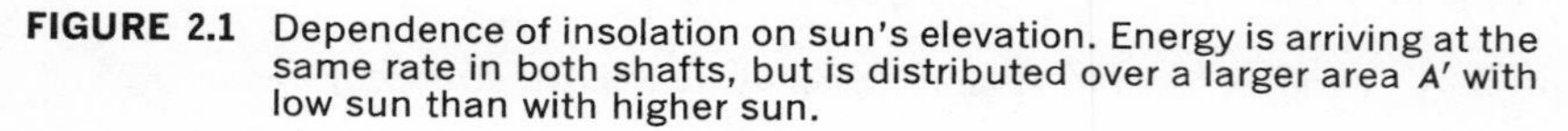

FIGURE 2.1 Dependence of insolation on sun's elevation. Energy is arriving at the same rate in both shafts, but is distributed over a larger area A' with low sun than with higher sun.

It is important to note that the solar constant is the same above any point on the earth. But the flux of energy intercepted by a unit of horizontal area at the surface of a round earth is not, even when atmospheric effects are neglected. The situation is as pictured in Figure 2.1. The areas A are unit areas perpendicular to the beam and intercept a given flux of solar energy. At lower solar angles, this amount of energy is distributed over the larger area A', so that the energy flux at the ground is reduced by an amount controlled by the angle h. The energy flux on a horizontal surface is called the insolation J.

How the simple fact that the earth is spherical influences the temperatures can be seen from Table 2.1, in which the average annual insolation values (neglecting the effect of the atmosphere) and the average annual air temperatures are given by latitudes on the northern hemisphere.

TABLE 2.1
Average Insolation and Temperature at Various Latitudes

LATITUDE DEG. NORTH	0	10	20	30	40	50	60	70	80	90
INSOLATION (kly/yr)*	321	317	304	282	254	220	183	152	138	133
TEMPERATURE (°C)	26.0	25.5	23.0	17.0	11.0	4.0	−3.0	−11.5	−19.5	−26.0

*kly/yr = kilolangley per year = 1000 calories per square centimeter per year

EXERCISE 3

Plot the data from Table 2.1 on graph of Figure 2.2. The insolation values given are those arriving at the top of the atmosphere. Within the atmosphere, certain processes deplete this insolation before it reaches the ground. How important are these processes in controlling the poleward variation of the mean annual temperature? Please note, that in this and other graphs, the scales of ordinates and abscissae have been selected so that trends of curves can be related to one another. No great significance should be attached to the intersection of curves and similar features of relative position of curves.

The greatest seasonal variation in insolation is found at the polar caps. For example, at the Arctic Circle ($66\frac{1}{2}$°N Lat.) on 22 December, there is essentially no insolation. On 22 June, the insolation J (neglecting the atmosphere) at noon can be calculated from the formula $J = J_0 \sin h$, where J_0 is the solar constant and h is the sun's altitude equal to 47° at that time.

EXERCISE 4

Calculate the insolation on 22 June at the Arctic Circle at noon, noting that the sine of 47° is 0.73. Compare the range of insolation observed at the Arctic Circle during the year with the result of Exercise 2. Which effect is responsible for the seasons, the varying distance of the earth from the sun or the varying solar angles at the earth's surface?

It is the variation through the year of the sun's elevation above the horizon, producing more intense insolation and longer days at certain times, that accounts for the seasons, not the slight effect due to the variation of the earth's distance from the sun. This being the case, it is not surprising to find that seasonal variations are much more pronounced at middle and high latitudes; inasmuch as the noonday sun is always high in the sky at low latitudes, seasonal effects are quite small there.

The conclusion is that because the earth's orbit is almost circular, seasonal effects would be very trivial if it were not for the inclination of the earth's axis of rotation to the plane of its orbit. This inclination is 66½°, or the angle between the earth's axis and the perpendicular to the orbital plane is 23½°. The axis is fixed relative to distant stars in the universe; its northern end points to the Pole Star. As the earth orbits around the sun, the sun's noon position appears to move from the equator, where it is in the zenith, that is, directly overhead, on the spring equinox, to a zenith position at the Tropic of Cancer (23½°N) on the northern summer solstice. From there it moves back toward the equator, where it is again overhead on the autumnal equinox. Then the sun appears to move to the southern hemisphere, where it is in the zenith at noon over the Tropic of Capricorn (23½°S) on the northern winter solstice, and so on. The latitude angle at which the sun is in the zenith at noon is called the sun's declination. These geometric aspects are shown graphically in Appendix I, Experiment 1, which should prove to be an aid in visualizing this concept.

Seasonal changes in the daily path of the sun across the sky produce two different effects that alter the daily insolation received at any given latitude. First, the angle at which the sun's rays impinge upon the ground varies as discussed above. Second, the length of daylight varies. The latitudinal distribution of these quantities is shown in Figure 2.3, for the summer and winter solstices. At the equator, every day is 12 hours long; during the northern hemisphere summer, every latitude poleward of the Arctic Circle has at least one 24-hour period of continuous sunlight, while every latitude poleward of the Antarctic Circle has at least a 24-hour period of continuous night at that time, which is winter in the

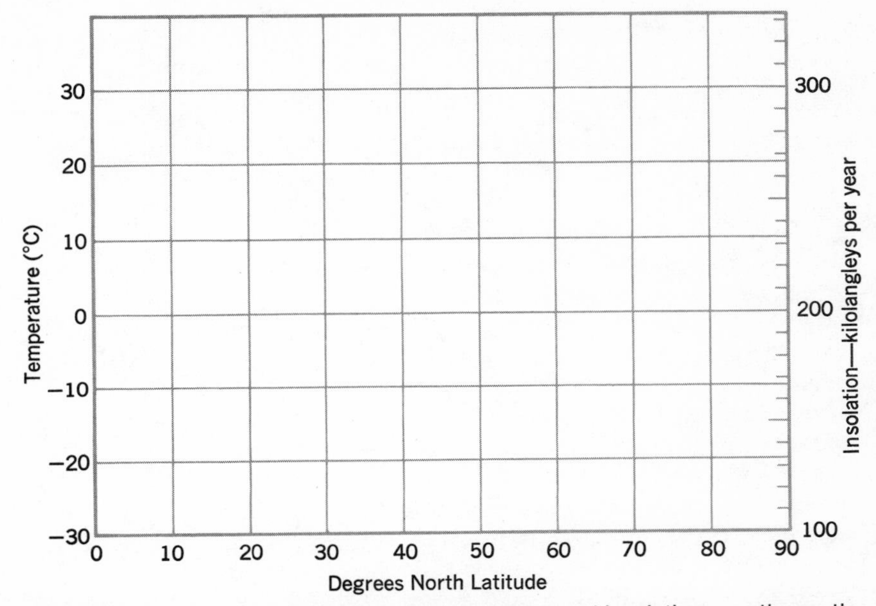

FIGURE 2.2 Latitudinal distribution of temperature and insolation over the northern hemisphere.

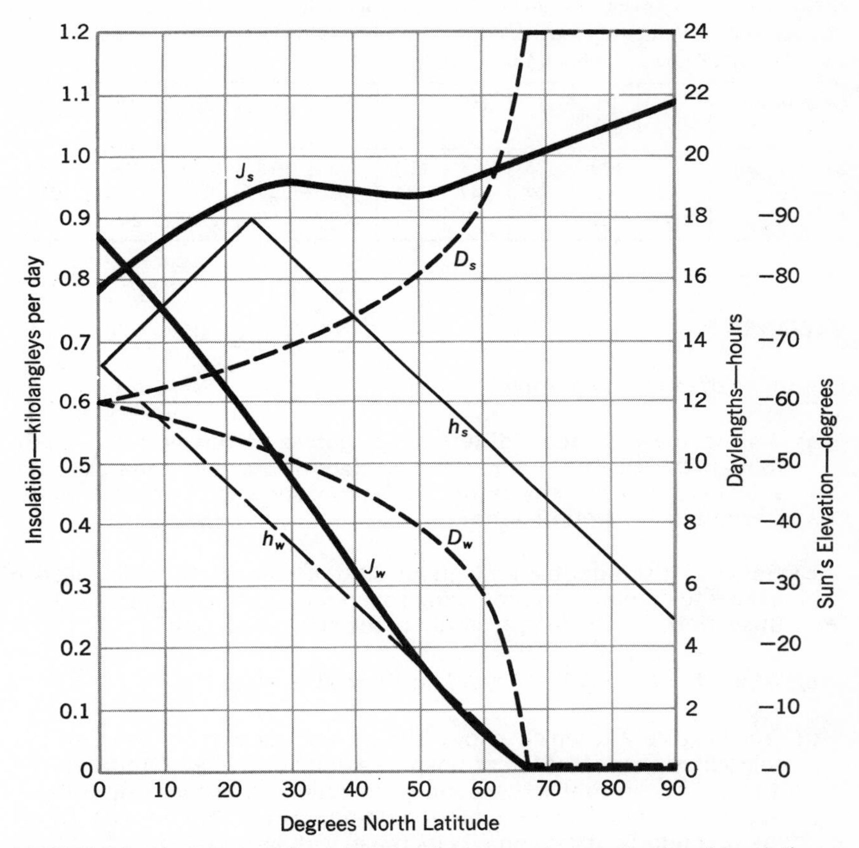

FIGURE 2.3 Maximum and minimum values at various latitudes of insolation J, daylength D, and elevation h of the sun at noon. Subscript s is for maximum on summer solstices; subscript w is for minimum on winter solstices.

southern hemisphere. Latitudes between the equator and $66\frac{1}{2}°$ are intermediate between the situations described; generally, the annual variation of daylength increases with latitude. As an interesting consequence of the long summer days at high latitudes, the greatest daily totals of energy are received over the poles, despite the relatively low elevation angles of the sun.

In order to consider the effects of varying sun's elevation and daylength separately, the differences between their maximum and minimum values are extracted from Figure 2.3 and listed in Table 2.2.

TABLE 2.2
Differences (Δ) between Maximum (22 June) and Minimum (22 December) Values of Insolation (*J*), Sun's Elevation (*h*), and Daylength (*D*) at Various Latitudes

Latitude Deg. N	0	10	20	30	40	50	60	70	80	90
ΔJ, (ly/day)	−70	120	305	485	620	765	915	1010	1050	1080
Δh (deg.)	0	20	40	47	47	47	47	$43\frac{1}{2}$	$33\frac{1}{2}$	$23\frac{1}{2}$
ΔD (hr)	0	1.2	2.4	3.8	5.6	8.4	13.0	24.0	24.0	24.0

EXERCISE 5

Plot these data on the graph of Figure 2.4.

a) Why is the insolation difference negative at the equator, even though the sun's elevation and daylength are the same on both solstices, that is, why is the insolation somewhat greater on 22 December than on 22 June?

b) Between the equator and about 30 degrees north, which effect seems to dominate in producing seasonal differences in insolation, the variation of daylength or that of the sun's elevation?

c) What about the latitude zone north of 60 degrees?

d) From Figure 2.2, which shows the strong correlation between insolation at the top of the atmosphere and surface air temperature, it can be seen that the seasonal insolation differences discussed above are an indication of seasonal temperature differences. Does the magnitude of seasonality increase with latitude?

e) Why would a large winter-to-summer temperature contrast occur at a place like Chicago (42°N)? Appraise this by comparing the insolation difference at 42°N in Figure 2.4 with the average insolation at the same latitude in Figure 2.2 (see Authors' Note for conversion).

There is, however, an important second-order effect on the insolation received at the ground which takes on greater importance at the higher latitudes. The pathlength through the atmosphere, which depletes the radiation, diminishes as the solar elevation increases. This pathlength is defined as being unity, that is, one atmospheric thickness, when the sun is in the zenith, and increases to about 40 thicknesses when the sun is at the horizon.

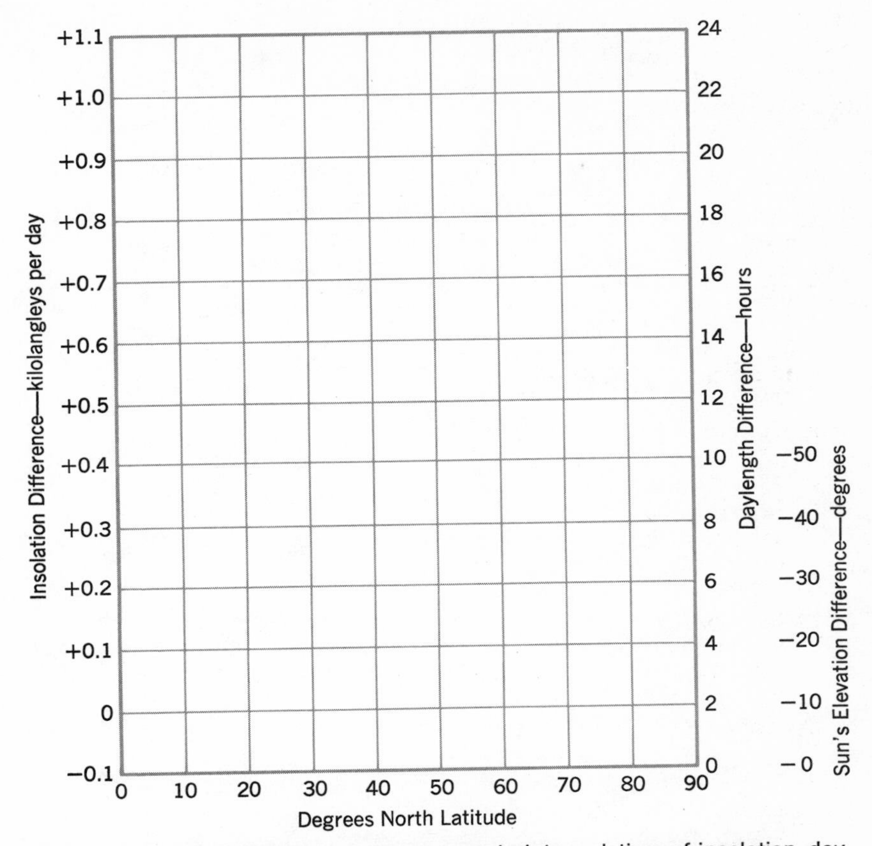

FIGURE 2.4 Differences between summer and winter solstices of insolation, day length, and sun's noon elevation at various latitudes.

2.2 DEPLETION OF SOLAR RADIATION BY THE ATMOSPHERE

In addition to the controls imposed by the earth–sun geometry, three processes modify or diminish solar energy on the way through the atmosphere. The first is scattering by the molecules of the atmospheric gases, smoke, dust, and the tiny droplets and ice crystals of which clouds and fog are composed. The scattering process redirects the sun's rays in all directions; in turn, the scattered rays are again scattered when they meet with other molecules or particulate suspensions. Because molecular scattering is strongest in the short wavelengths of sunlight (blue and ultraviolet), the clear sky looks blue in daytime; but thin clouds, haze, fog, or smog look essentially white, for the scattering particles are then much larger than molecules, and larger particles scatter all wavelengths about equally.

EXERCISE 6

a) When we fly to high altitudes, does the sky above us get brighter or darker?

b) Lighter blue, or deeper blue? Explain.

The scattered portion of solar energy is not entirely lost to the earth's surface; about half of the scattered energy is directed earthward and half of it toward space. The portion scattered earthward is called sky radiation, which represents a substantial fraction of the total radiant energy received by the earth. As a matter of fact, in polar regions, where the sun does not rise above the horizon for part of the winter, sky radiation is the sole source of incoming radiative energy. Even in lower latitudes, the sky radiation may exceed the direct insolation when the sun is at low elevation angles. Of course, when the sky is overcast with dense clouds, only sky radiation reaches the earth, regardless of latitude or sun's elevation. The total flux of energy from both sun and sky on a horizontal surface is called global irradiance. In Table 2.3 the average percent contribution of sky radiation to global irradiance is given together with cloudiness for various latitude zones in the northern hemisphere.

At low latitudes where the solar angle varies little during the year, the sky-radiation contribution increases as the cloudiness does from winter to summer. By contrast, the contribution by sky radiation decreases from winter to summer at high latitudes, although cloudiness increases even more there.

TABLE 2.3
Sky Radiation as a Percentage of Global Irradiance.
Percent Cloudiness Is Given in Parentheses.

Zones Deg. N	Winter Solstice	Equinoxes	Summer Solstice
0–30	38 (37)	39 (42)	42 (48)
30–60	56 (55)	45 (48)	44 (48)
60–90	100 (49)	71 (65)	62 (75)

EXERCISE 7

a) Why does the relative contribution of sky radiation at high latitudes decrease from winter to summer?

b) Why is the increase in sky-radiation contribution with latitude greatest during the winter solstice?

Reflection is another process by which solar energy is depleted. The fraction of the sun's radiation that is reflected is called the albedo. The albedo of various types of surfaces on the earth varies over a wide range as can be seen from Figure 2.5. For the earth's surface and the atmo-

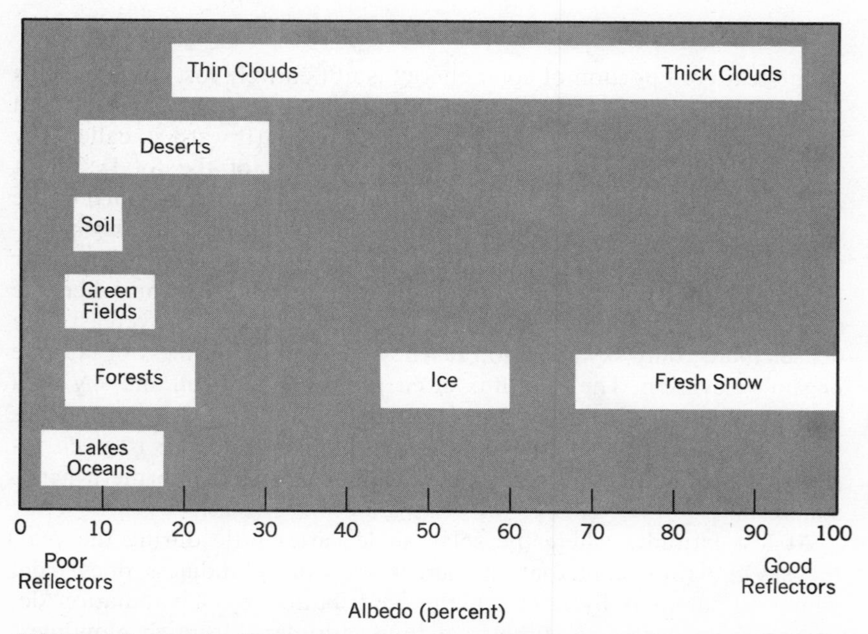

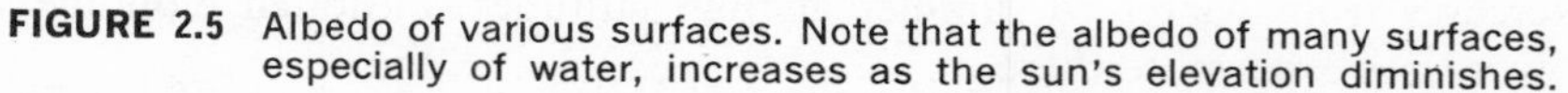

FIGURE 2.5 Albedo of various surfaces. Note that the albedo of many surfaces, especially of water, increases as the sun's elevation diminishes.

sphere, taken together, the mean albedo is approximately 35 percent. However, the albedo varies considerably between the equator and the poles, largely because of the snow and ice fields and greater cloudiness in the polar regions. See Figure 3.4 which shows the greater albedo near the poles. In general, the earth is a bright planet; compared to the brightness of a moonlit night on earth, an earthlit night on the moon is quite spectacular inasmuch as the earth's albedo is about five times as large as that of the moon. (See Figure 2.6.)

FIGURE 2.6 Earth view from Apollo 8 over Lunar surface. (Photo courtesy of NASA.)

The albedo of deep clouds is greatest, hence thick clouds are very bright clouds as seen from above, but very dark as seen from beneath. As much as 90 percent of the incoming energy may be reflected back into space by thunderclouds. Because they reflect so well, clouds and fog represent important controls of the energy that may penetrate to the earth's surface. Table 2.4 gives the mean cloudiness (in percent of the sky area) at various latitude belts.

TABLE 2.4
Mean Cloudiness (percent) by Latitude Zones

Latitude Deg. N.	0–10	10–20	20–30	30–40	40–50	50–60	60–70	70–80	80–90
Cloudiness	52	47	43	46	53	60	65	66	62

EXERCISE 8

a) Does the albedo of the ground plus atmosphere increase or decrease poleward? Why?

b) What implications does this have for the air temperature difference between the equator and the poles?

Absorption by atmospheric gases and suspensions removes less than 20 percent of the incoming radiant energy, and only certain wavelengths are absorbed. Whenever absorption occurs, the temperature of the air rises. The biologically lethal portion of ultraviolet is absorbed by ozone in a layer that is found between 20 and 50 km heights. Almost no visible light is absorbed in the atmosphere; but the sun's energy in the longer infrared wavelengths is absorbed quite efficiently by water vapor, clouds, and carbon dioxide. The amount of energy absorbed by dust and smoke is usually very small and, as a transient phenomenon, of little importance. However, when catastrophic volcanic eruptions, such as those of Krakatau (1883; 6°S, 105°E) or Katmai (1912) in Alaska fill the atmosphere with vast amounts of persistent dust, insolation at the surface of the earth may be markedly reduced. In the cases mentioned, the reduction amounted to 20–30 percent at many places. Ahead of his time as he was in many of his pursuits, Benjamin Franklin was the first to suggest that volcanic dust could produce climatic changes by altering absorption.

The total reduction of insolation by scattering, reflection, and absorption is called depletion. Although depletion by the atmosphere is substantial, scattered sky radiation tends to compensate for it so that the variations in radiant energy at the earth's surface due to atmospheric conditions are, on the average, small in comparison to those produced by the varying solar angle. In Table 2.5 the percent depletion of direct insolation is given for three latitude belts under average turbidity and cloudiness conditions on the summer and winter solstices and the equinoxes.

TABLE 2.5
Depletion (percent) of Insolation by the Atmosphere

Zones Deg. N	Winter Solstice	Equinoxes	Summer Solstice
0–30	67	68	71
30–60	76	69	72
60–90	—	81	81

EXERCISE 9

a) Why is the depletion greater at high latitudes?

b) Why are the differences in depletion between the equatorial and the middle latitudes greatest on the winter solstice?

The data of Table 2.5 show that the depletion by the atmosphere is quite large under all conditions. Despite this, about half of the insolation arriving at the top of the atmosphere actually reaches the ground. This apparent discrepancy can be attributed to the fact that some of the depleted energy is scattered downward in the form of sky radiation. In summary, the earth-atmosphere system as a whole absorbs about 65 percent of the incoming radiative energy — the albedo being 35 percent — but approximately 50 percent of it actually reaches the earth's surface, the difference (15 percent) between the two percentages being absorbed by the atmosphere. However large, depletion does not materially alter the seasonal and latitudinal control on temperature exerted by insolation at the top of the atmosphere.

2.3 TERRESTRIAL RADIATION AND COUNTERRADIATION

The sun, being hot, radiates about 45 percent of its energy as visible light, the remainder in the ultraviolet and shorter wavelengths at one end of the electromagnetic spectrum, and in the infrared and longer wavelengths at the other end. Although all bodies radiate, cooler ones like the earth radiate only long-wave energy. The water vapor and carbon dioxide in the atmosphere absorb this terrestrial radiation very well; the absorbed energy is reradiated both upward and downward by these gases, the downward component being called counterradiation. By this process, the heat loss by terrestrial radiation is slowed down, because the atmosphere permits the passage of sunlight to the ground, but intercepts much of the energy going out from the ground. This is an important mechanism for trapping warmth in the lowest part of the atmosphere, which transforms the planet from a barely habitable one for life as we know it to one that is quite suitable.

Inasmuch as water vapor and clouds are largely responsible for infrared absorption, clear dry localities tend to have large temperature decreases at night and large increases in daytime, while moist cloudy ones do not. The absence of much counterradiation helps to explain why high mountains that extend above most of the water vapor blanket, half of which lies below 2000 meters height, are cold.

2.4 THERMAL PROPERTIES OF DIFFERENT SURFACES

Even when the incoming radiation is the same, the various surfaces utilize the finally absorbed energy in different ways. Bare soil, which is opaque to sunlight, absorbs all the available energy in a thin top layer and therefore gets warm quickly. In the absence of sunlight, the energy stored in that thin layer is readily radiated back into the atmosphere; therefore bare soil cools rapidly. When there is a cover of vegetation on the ground, part of the incoming energy is used in the conversion of radiant into chemical energy, that is, photosynthesis, and part of it is used to evaporate water from the plant surfaces (transpiration). The net effect is a lesser temperature rise of vegetated than of bare ground surfaces. (See Appendix I, Experiments 2 and 3.) Snow and ice surfaces never get "warm," partly because of their high albedo, but also because they lose energy in the infrared very readily. Further, part of any small amount of absorbed energy is utilized in evaporating the water substance and part in raising the temperature to the melting point, at which temperature a large amount of energy is used to melt the ice. No further temperature rise can occur until all the ice is melted.

Locally, the presence or absence of snow cover greatly alters the rate at which soil gives up heat to the atmosphere. Loose, fresh snow is a very good insulator; a snow cover of about two feet will inhibit frost penetration entirely. Here is one of many examples of a climatic effect sustaining itself. Bare soils in cold climates freeze and, because of the ice, become even better heat conductors, thus promoting still more rapid heat loss and further frost penetration. By contrast, a deep snow cover established early in winter may virtually prevent significant frost penetration into the ground. Cold snow-free winters may cause some soils to freeze to a depth of several feet; this can become very significant for the survival or frost-killing of certain plant roots and bulbs. Furthermore, spring runoff from frozen ground can produce flooding. In this connection, the removal of frost depends more on the percolation of warm spring rains and snowmelt than on the absorption of heat from the sun.

Temperature changes of free water surfaces are also quite slow as compared to soils. Not only does sunlight penetrate deeply into the water and thus distribute energy over a large volume of water, but a substantial portion is used to evaporate water into the air above it. Furthermore, the specific heat of water, that is, the amount of heat required to raise the temperature of a unit mass of substance by 1°C, is about four times as great as that of soil, so that the same radiant energy will heat a given mass of soil to a higher degree than an equal mass of water. Most importantly, however, water is mobile so that convection currents in the water, the flow of the water itself, and wave action will mix warmer with colder water. The water is kept cool as compared to soil under a given condition of incoming radiation by all these factors, but mainly because of the greater volume of water involved.

As far as outgoing radiation is concerned, both water and soil surfaces are equally efficient radiators, but water, having easily accessible heat stored at greater depths, cools much more slowly than does soil under the same conditions.

3 energy exchange

3.1 REGIMES OF ATMOSPHERIC CIRCULATION

It is an interesting fact that a rotating fluid, like the atmosphere, under unequal heating, or just as importantly, unequal cooling, will set itself into motion in patterns that are organized and, to a certain degree, predictable. Experiments have shown this behavior to be reproducible by models, and it is known from study of these models that such quantities as the rate of rotation and the horizontal temperature differences are decisive for the wind pattern to be expected. However, many random turbulent motions, both large and small, become superimposed upon the organized flows and often nearly obscure them, making it necessary to examine first the average wind patterns in order to understand the effects of organized motions, and then the random motions separately to understand their effects.

The winds near the earth's surface are schematically displayed in Figure 3.1. The existence of wind patterns on each hemisphere is immediately apparent. The easterly[1] surface winds at low latitudes are remarkably regular, so that stations in the tropics exhibit highly persistent wind directions. Irregularities that do occur are associated with occasional precipitation events. By contrast, the winds at higher latitudes are much less regular, and a mean westerly current only appears when averaged over longer time periods. At all latitudes the persistence of wind is greater over the oceans than over land; however, where mountain ranges channel the air flow into preferred directions, stations well outside the tropics and away from oceans might be said to have very persistent winds also. In Table 3.1 the average frequencies of various wind directions (omitting calms) are given for three stations: two in middle latitudes and one in the Trade wind zone. The data for Honolulu are plotted on Figure 3.2, which is called a wind rose.

[1]The wind direction is that from which the wind blows; for example, an east wind blows from east to west.

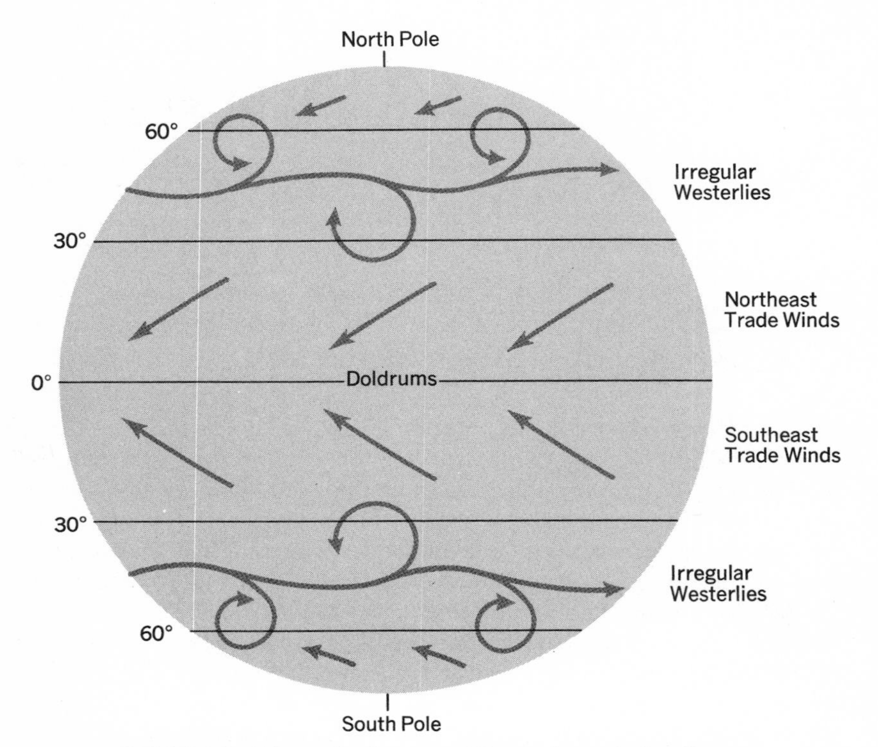

FIGURE 3.1 Schematic diagram of the surface-wind patterns.

TABLE 3.1
Mean Annual Percent Frequencies of Wind Directions in Different Regimes

Station	Wind Directions							
	N	NE	E	SE	S	SW	W	NW
Honolulu, Hawaii	8	46	29	4	4	2	2	5
Boise, Idaho	7	3	5	33	10	3	12	27
Indianapolis, Indiana	10	10	10	12	14	19	12	13

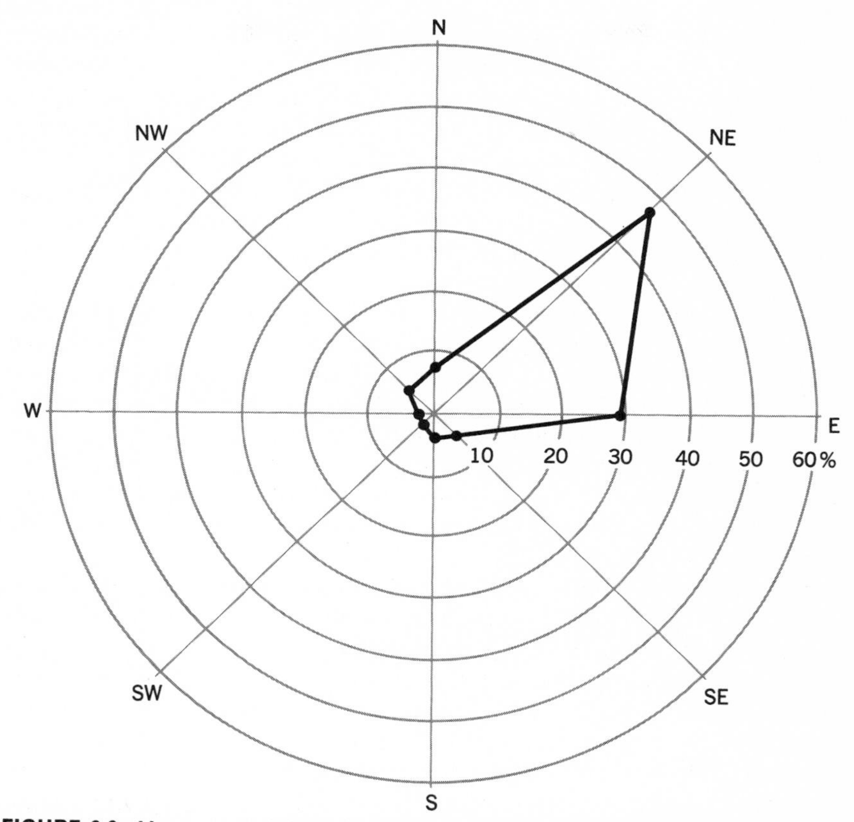

FIGURE 3.2 Mean annual frequencies of various wind directions at Honolulu, Hawaii.

EXERCISE 10

Plot the wind roses for Boise and Indianapolis on the diagram of Figure 3.3.

a) If you did not know that Boise, Idaho, lies between mountains, what characteristics of the wind rose would enable you to differentiate between the situation there and that at Honolulu?

b) For Indianapolis add the frequencies of the three westerly components and do the same for the three easterly components. Determine by what percentage the frequency sum of the one component exceeds that of the other component sum.

c) Do the same for the frequency data of Honolulu; are the Trade winds more regular than the so-called "prevailing westerlies"?

The approximate structure of the surface-wind patterns has been known since the eighteenth century and was supposed to consist of three circulation cells on each hemisphere. Most of the theories that were developed to explain the general circulation of the atmosphere attempted to show how these three cells could transfer energy received from the sun in the tropics toward the higher latitudes to drive the strong winds observed there. Little was known about the winds at higher levels, but these were supposed to be reverse flows maintaining the circulation inasmuch as the air cannot accumulate in any one region over long periods of time. In 1735 the English scientist George Hadley postulated a large-scale circulation that relied for a driving mechanism on the rising of heated air near the equator and its outflow aloft toward the poles, sinking at higher latitudes, and a return flow at lower levels toward the equator. This formulation was in agreement with what could be observed in the laboratory and, when the effects of the earth's rotation were included, seemed to explain the northeasterly Trade winds observed by Edmund Halley some years earlier. Further observations led to modification of the theory, the principal ones adapting to the three-cell structure.

Since the advent of the airplane and balloon-borne sounding devices, observations of winds aloft have considerably improved with the result that the three-cell model has been found to be inadequate in explaining the exchange of energy between the equator and the poles. In particular, the mean north–south, or meridional, circulations outside the tropics are known to be very weak and to play only a minor role in the poleward exchange of energy and water vapor. The middle-latitude regime is very vigorous and capable of large exchanges to and from the poles, but these are accomplished by transitory circulation features, rather than by a mean meridional circulation.

3.2 THE VARIABLE WINDS OF MIDDLE AND HIGH LATITUDES

What the upper-air observations have shown is that, on the average, the middle latitudes of both hemispheres exhibit westerly winds which increase in speed with height up to the top of the troposphere, called tropopause, usually found at heights between 8 and 15 km. But the energy of these winds at higher latitudes, being supplied near the equator by the sun, cannot be carried northward by any simple mean south–north circulation. In other words, west winds cannot carry anything northward or southward.

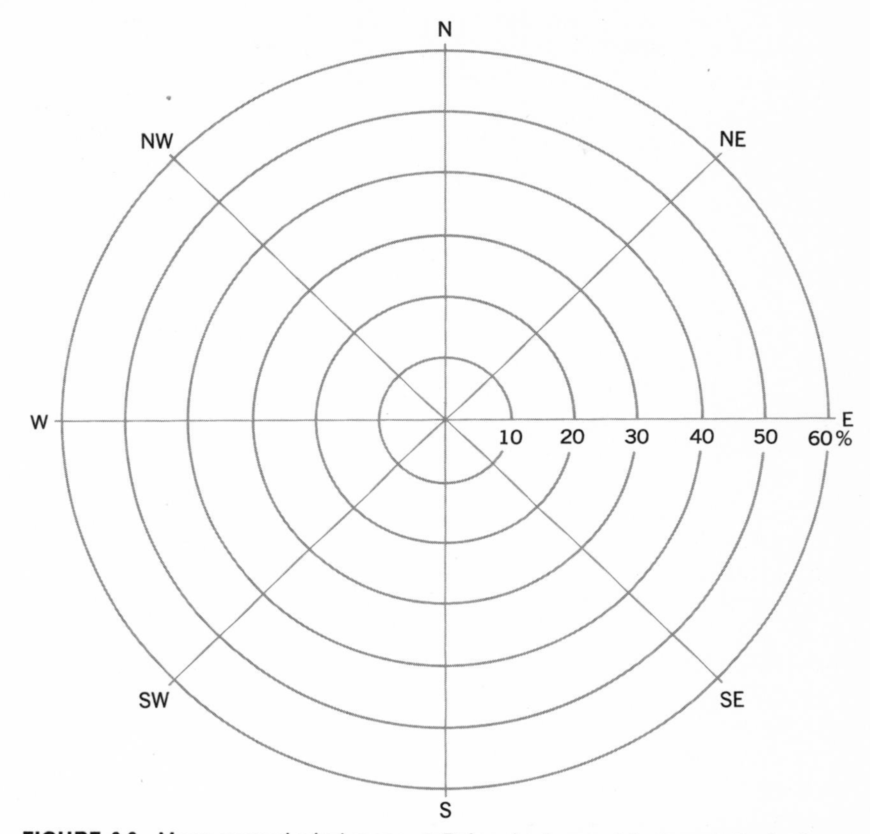

FIGURE 3.3 Mean annual wind roses at Boise, Idaho, and Indianapolis, Indiana.

EXERCISE 11

The energy of dry air is the sum of three components: potential energy, which is proportional to the height of the air above the ground, internal energy, which is proportional to its temperature, and kinetic energy, which is proportional to the speed of the wind. The potential and internal energies have been found to be positively correlated with each other. The kinetic energy of the air is greatest in middle latitudes where the strongest winds occur on the average. Would you expect the other two energy components to be a maximum at low, middle, or high latitudes?

It can be concluded that the atmosphere absorbs most of its energy in the tropics where it appears as internal and potential energy. The transport of energy thus acquired, which drives the strong winds found farther poleward, is largely accomplished by the large vortices or eddies

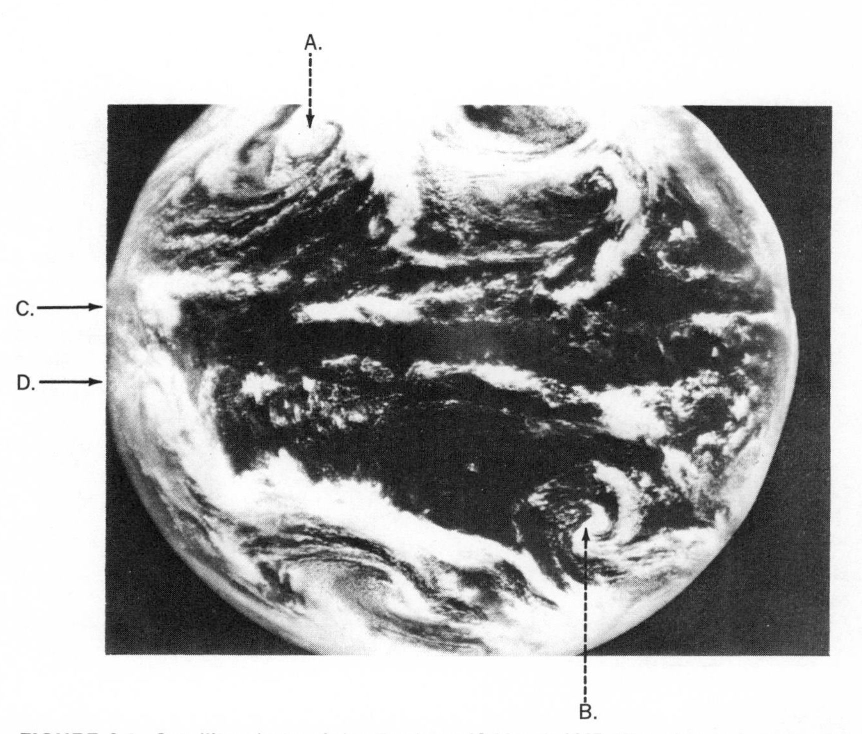

FIGURE 3.4 Satellite photo of the Earth on 28 March 1967. The picture is centered on a point on the equator in the middle of the Pacific Ocean. Letter A on upper left shows counterclockwise spiral of clouds associated with a low-pressure system on the northern hemisphere. Letter B on lower right shows clockwise spiral of clouds associated with low-pressure system on southern hemisphere. Arrows C and D point to lines of clouds associated with the Intertropical Convergence Zones on the northern and southern hemisphere, respectively. (Photo courtesy of NASA.)

in the middle latitudes. These vortices appear on weather maps as low- and high-pressure systems, called cyclones and anticyclones, respectively. Figure 3.4 is a typical satellite picture showing cloud patterns associated with cyclones whose rotation is counterclockwise on the northern hemisphere, clockwise on the southern hemisphere. The cyclones are normally regions of "bad" weather, and the clouds display the cyclonic circulation rather well. Individual vortices usually have lifetimes of the order of several days during which they develop, migrate, and dissipate. Using the northern hemisphere as an example, their importance in terms of energy transport lies in the fact that their transitory rotational circulations include north winds carrying air of low energy southward and south winds carrying air of high energy northward. However, they produce no net meridional circulation because the north winds cancel out the south winds. This is a very important point because it shows that the poleward energy exchange can be carried out without a mean poleward circulation.

EXERCISE 12

The net poleward heat transport across a particular latitude circle can be calculated as follows: The speed of the north–south components of the wind is multiplied by the temperature departure from the average for that latitude at numerous points (longitudes) along the latitude circle. Recall that the product of a negative number and a positive one is negative, and products of numbers with like signs are positive. These products are then added algebraically. Table 3.2 contains a grossly simplified set of observations for latitude 40°N where the mean temperature is 11°C; the + signs before the wind speeds indicate components from the south, — signs indicate components from the north, so that + means poleward transport, — means equatorward transport.

TABLE 3.2
Longitudinal Distribution of the Speeds of Wind Components from the South (+) and from the North (−) and Temperature Departures from the 40°N-latitude Average

LONGITUDE	0	40E	80E	120E	160E	160W	120W	80W	40W
WIND SPEED meters/second (mps)	+2	+1	+1	+3	−6	+3	−1	+1	−4
TEMPERATURE DEPARTURE (°C)	+5	+8	−3	+2	−6	+1	−4	0	−2
PRODUCTS:									

a) Enter the products of each pair of values in Table 3.2 and then add up all the products; the result, while not the actual transport of energy, is proportional to it.

b) What is the mean meridional wind speed across latitude 40°N?

c) How do the observations in Table 3.2 show that north winds usually are cold and therefore have low energy, whereas south winds are usually warm and therefore have high energy? This fact is essential for poleward exchanges in the absence of significant mean meridional circulations.

In middle and high latitudes, the winds are quite variable with respect to both direction and speed, but west winds occur with sufficient frequency and intensity to make the average flow westerly at the surface and to a considerable height above it. A minor exception occurs at low levels near the poles, where weak easterly winds are slightly more common, especially in summer. Because changing winds transport air masses from different regions, the climates in these latitudes are characterized by relatively large day-to-day changes in cloudiness, temperature, and water-vapor content. Nevertheless, some regularity must be present, because the periodic variation of insolation exerts a strong control on the mean temperatures, while the weather systems with their associated cloud and precipitation patterns superimpose their influences on a daily basis. After all, January in the United States is colder than July, but one given January day may be drastically colder than the preceding day.

The major feature that emerges in any discussion of the circulation at middle and high latitudes is the frequent formation of migratory cyclones and anticyclones. For this reason, some mention must be made of the favored cyclone, or storm, regions; on both hemispheres, the belt between about 45 and 60 degrees latitudes has the most frequent cyclones, because here the temperature contrasts are greatest on the average. Again, there is considerable variation in an east–west direction; on the northern hemisphere, the Gulf of Alaska and the northern portion of the North Atlantic are major cyclone centers with enough storms to produce the often-mentioned low-pressure "belt" of moderately high latitudes.

3.3 THE TROPICAL CIRCULATION CELL

Only in the tropics can one find the more regular, cell-like circulation suggested by Hadley with upward motion near the equator, outflow aloft, and sinking motions near latitude 30 degrees forming the subtropical high-pressure belts. A return flow called the Trade winds completes the circulation. Appropriately enough, the tropical wind regime is called a Hadley cell, one cell over each hemisphere. Even here, however, the situation is far more complicated than what is adumbrated above.

First, all of those upward, downward, and poleward motions are slow, on the average, and are superimposed on the easterly Trade winds near the surface and on a not inconsiderable westerly flow aloft. Second, the tropical Hadley cells tend to migrate northward and southward, seasonally with the sun, but lagging a month or two behind it. Thus, the Hadley regime is usually considered to obtain between latitudes 30° north and 30° south with some expansion poleward in the summer hemisphere. Until recently, it was believed that the northeast Trade winds of the northern hemisphere and the southeast Trade winds of the southern hemisphere converged into a single belt of very light, variable winds, called Doldrums, so that the air there is forced upward, causing a maximum of cloudiness and precipitation near the equator.

It should be noted that this is more than a casual statement of fact derived from observation; upward motion is at the roots of nearly all clouds and every significant precipitation event observed in the atmosphere. Briefly, the rising air expands and therefore cools. As it cools, the amount of water vapor that can be mixed with it diminishes so that some of the vapor condenses out. Conversely, subsiding air is compressed, its temperature rises, and therefore its capacity for holding water in vapor form is increased, so that any clouds present will dissipate.

The convergence of air at low levels and the consequent upward motion of air cause an accumulation aloft which leads to an outflow toward the subtropical high-pressure belts. The region in which this convergence occurs is called the Intertropical Convergence Zone, variously abbreviated as ITC, ITZ, or ICZ. The ICZ also tends to migrate northward and southward with the sun with some lag. However, some recent evidence suggests that the ICZ may not cross the equator but rather may have a double structure with one band of cloudiness and precipitation north and one south of the equator, and a relatively clear belt of presumably subsiding air very near the equator. The apparent motions of the ICZ may be associated partly with changes in intensity of the two belts rather than with actual movement, although some migration does seem to occur. See Figure 3.4 for a typical configuration of the ICZ with the cloudiness belt near 10°N somewhat better developed than that on the southern hemisphere. It should also be noted that neither of the convergence zones is a continuous belt around the earth. The migration of the ICZ is at least partially related to the monsoons of Asia, Africa, and Australia. In addition, the subtropical high-pressure belts with their attendant sinking air and clear skies migrate poleward as far as 40 degrees north in August and September of a normal year, virtually guaranteeing sunny summer weather to Greece, Italy, and California. Furthermore, the Hadley cells themselves contain pulsating and meandering high-pressure systems near their poleward borders which alternately distort, strengthen, and weaken the Trade-wind belts around the earth. The so-called Bermuda and Pacific Highs are examples; south of them the Trade winds tend to be more regular; at their western and eastern extremities, such as the region around the Gulf of Mexico, heat and moisture exchange with higher latitudes is enhanced.

Finally, there are day-to-day variations in the tropics, not as pronounced as those at higher latitudes, but usually associated with periods of increased precipitation in response to relatively minor variations in the wind patterns. The tropical Hadley cells are possibly structured as depicted schematically in Figure 3.5, but some of the detail is still unknown.

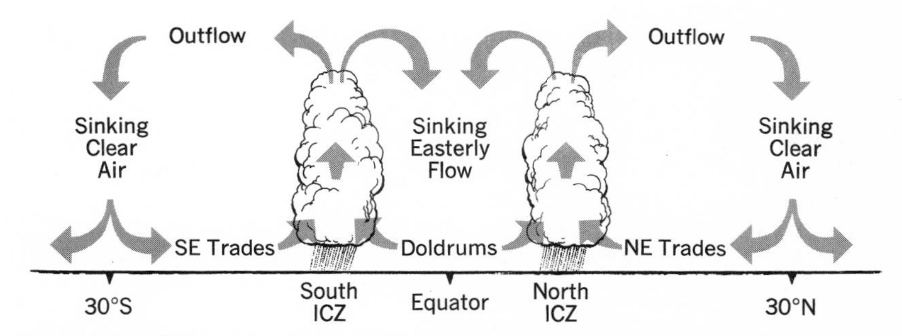

FIGURE 3.5 Schematic diagram of a tentative hypothesis concerning the structure of the Hadley Cell in the Tropics.

In summary, the imbalance of heating between the equator and the poles is modified by poleward heat transport. This is accomplished in the mean meridional circulation of the tropical Hadley cell and by the transient cyclones and anticyclones of middle latitudes. These cyclonic and anticyclonic eddies, though entirely obscured by west to east motions in the averages, must form in response to the horizontal temperature differences of middle latitudes. In terms of energy, the temperature differences arise both from the solar input to the tropical atmosphere and the radiational heat losses from polar regions. Thus, the atmospheric circulation runs on, produced by the very temperature contrasts it is constantly attempting to destroy.

4 the role of water

4.1 SOME PROPERTIES OF WATER AND ITS WORLDWIDE DISTRIBUTION

The presence or absence of water either as a solid or liquid at the earth's surface or as vapor in the atmosphere exerts a great influence on the observed climate. Also, the manner and rate of augmentation of the local water supply by precipitation are obviously important climatic components. Some of the effects produced by water are immediately evident, others are quite subtle but no less important. For example, an average annual snowfall of 575 inches at Paradise Ranger Station, Washington, makes this place one of the snowiest on earth; less obvious is the large amount of heat released in the atmosphere by the condensation–precipitation process.

About 99 percent of the earth's presently available water is in the liquid state and 97 percent of that amount is found in the oceans. Most of the rest is locked up in the great glaciers over Antarctica and Greenland; atmospheric storage of vapor is trivial by comparison. It has been calculated that enough new, or juvenile, water emanates from the bowels of the earth through volcanoes and is released from crystals to raise the level of the seas one meter every million years, but losses to space undoubtedly reduce this. In any case, the quantity of water on earth is quite constant over long periods of time.

Inasmuch as continents only occupy slightly more than one fourth of the earth, a liquid water surface constitutes the lower boundary of the atmosphere at most places; hence, the characteristics of a water surface do much to determine the earth's climate. Indeed, the distinction between maritime and continental climates is one of the simplest and most meaningful encountered in climatology. In this context, because they evaporate water into the air at a very low rate, glaciers, ice caps, and snowfields can be viewed as cold continents; in fact, some oceanic points at high latitudes exhibit continental climates in winter and maritime climates in summer. Coastal regions may also have alternating maritime and continental climates during the year as a result of wind shifts from

onshore to offshore flow. The usual case, however, is one in which a large water surface is an omnipresent control on the climate over and near it, reducing the variation of temperature, providing a water reservoir for precipitation and fog, as well as presenting a smooth surface to an unobstructed wind. Indeed, maritime locations are usually windier than inland points. Despite Chicago's nickname, Boston is the windiest major city in the contiguous United States, although Chicago probably got its reputation from the strong breeze off Lake Michigan.

Large bodies of water are very efficient heat storage facilities mainly because of the mobility of water. But the suppression of temperature variability at maritime locations is not entirely due to the buffering effect of heat storage in the ocean. In the atmosphere, the gas water vapor itself absorbs radiant heat, and cloudiness also blankets the earth thereby lowering the maximum temperatures and raising the minimum temperatures that would otherwise be observed. Clouds reduce the daily range of temperature by trapping the earth's heat close to the ground at night and by reflecting sunlight in daytime. Water vapor, by reradiating part of the absorbed heat back to the surface, is a rather effective blanket at night, whereas this effect is relatively unimportant in daytime. As an example, observations made at Jacksonville, Florida, in June 1964 are given; the nights of both periods shown in Table 4.1 were cloud-free with light, variable winds. The maximum temperatures during the afternoons indicate that the air was similar in its thermal characteristics for both periods. Tabulated with the temperature is the dew point, which is the temperature to which the air must be cooled in order to become saturated. The dew point is a measure of the amount of water vapor; the higher the dew point, the greater the amount of water vapor in the air.

TABLE 4.1
Diurnal Variations of Temperature and Dew Point at Jacksonville, Fla.

JUNE 1964	TIME (hours)	TEMP. (°F)	DEW PT. (°F)	JUNE 1964	TIME (hours)	TEMP. (°F)	DEW PT. (°F)
3	1400	90	58	22	1400	90	68
	1800	84	58		1800	87	71
	2200	76	65		2200	80	73
4	0200	66	64	23	0200	78	71
	0600	65	63		0600	73	70
	1000	83	62		1000	85	68
	1400	86	60		1400	89	71

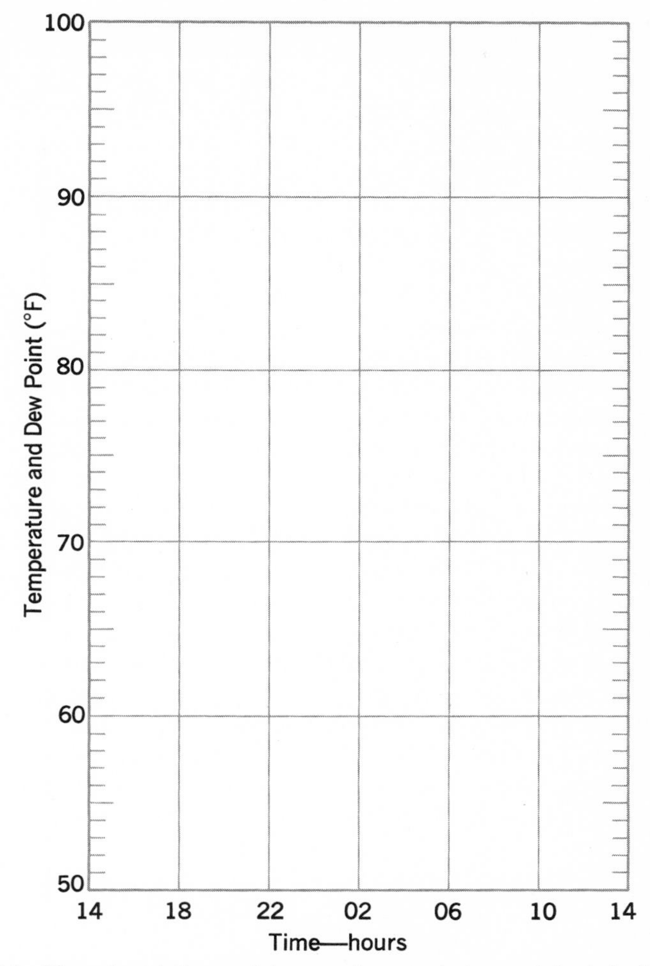

FIGURE 4.1 Diurnal variations of temperature and dew point at Jacksonville, Florida, on 3/4 June and 22/23 June 1964.

EXERCISE 13

Plot the data on the graph of Figure 4.1.

a) On which night did the air contain less water vapor?

b) How did the water vapor affect the minimum temperature?

c) The maximum temperature?

d) The range of temperature?

The amount of water vapor residing in the atmosphere can be expressed in terms other than dew point, such as by relative humidity or by precipitable water. Relative humidity is simply the percentage of saturation at a given temperature. It varies inversely as the temperature, when the dew point remains constant, which means that the relative humidity can vary although the amount of water vapor does not. Somewhat more easily visualized is the precipitable water which is the depth of liquid water that would result at the surface of the earth, if all the vapor in the atmosphere over the particular region were condensed out and fell as rain (an impossible proposition). The mean annual value of precipitable water for the whole earth is about one inch (2.5 cm).

EXERCISE 14

The mean annual precipitation for the whole earth is about 40 inches (100 cm).[1] How long does the average water-vapor molecule remain in the air between the time it evaporates from the surface and the time it precipitates again?

This exercise indicates that there is a rather rapid cyclic process continually producing evaporation, condensation in the air, and precipitation back to the ground; these processes together with runoff on the surface in the form of rivers, as well as storage underground, are called the hydrologic cycle. The main source of water vapor is evaporation from the subtropical oceans. This is true because warm water evaporates more rapidly than does cold water, other conditions being the same. This fact is demonstrated in Figure 4.2; the evaporation is much greater from the warm water at lower latitudes than from the cold water at higher latitudes. However, evaporation also depends on the dryness of the air above the water surface and the effectiveness of the wind in producing

[1]This value is still rather uncertain; the difficulties of obtaining reliable rainfall measurements over the oceans have been insurmountable to date.

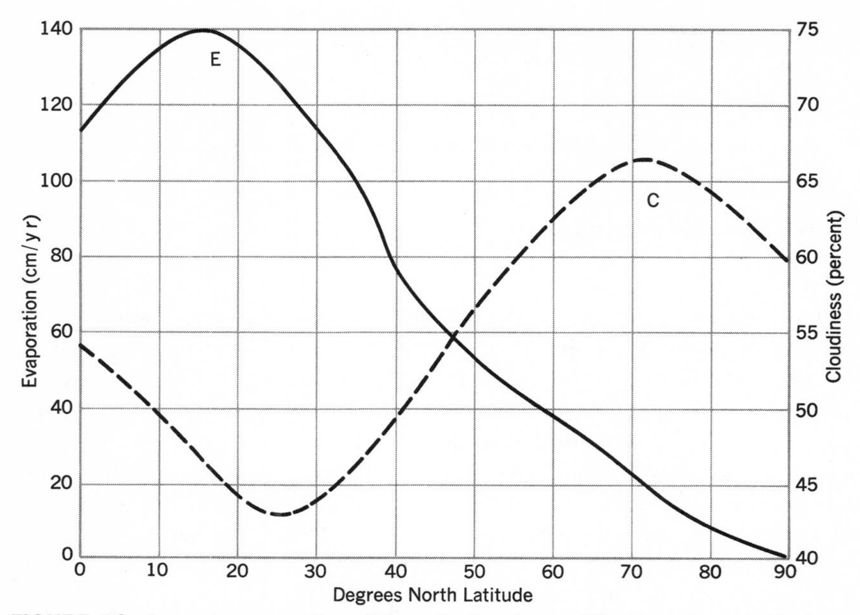

FIGURE 4.2 Annual evaporation (E) and cloudiness (C) at various latitudes, northern hemisphere. (After W. D. Sellers)

vertical mixing by means of eddies that transport humidified air upward and drier air from aloft downward. Nevertheless, the dryness of the air and the wind are secondary to surface temperatures in controlling evaporation rates.

The condensation portion of the hydrologic cycle may occur at a considerable distance from the region where the water had evaporated. Inspection of Figure 4.2 shows that the mean annual cloudiness is greatest at high latitudes and smallest in the subtropics, whereas evaporation is greatest in the subtropics and very small at high latitudes. In other words, evaporation proceeds most rapidly not only where the ocean surfaces are warm, but also where dry air is sinking from aloft. Inasmuch as the evaporation process at low latitudes consumes heat, some of which is released in the condensation process at high latitudes, this aspect of the hydrologic cycle involves a transport of "hidden," or latent, heat which helps to mitigate the temperature difference between the equatorial and polar regions. In general, latent heat released by condensation is of major importance in atmospheric energy transactions.

In addition, the amount of water vapor or precipitable water present in the atmosphere over any given place exerts a strong, but not decisive, control on the observed amounts of precipitation. Heavy rains can only

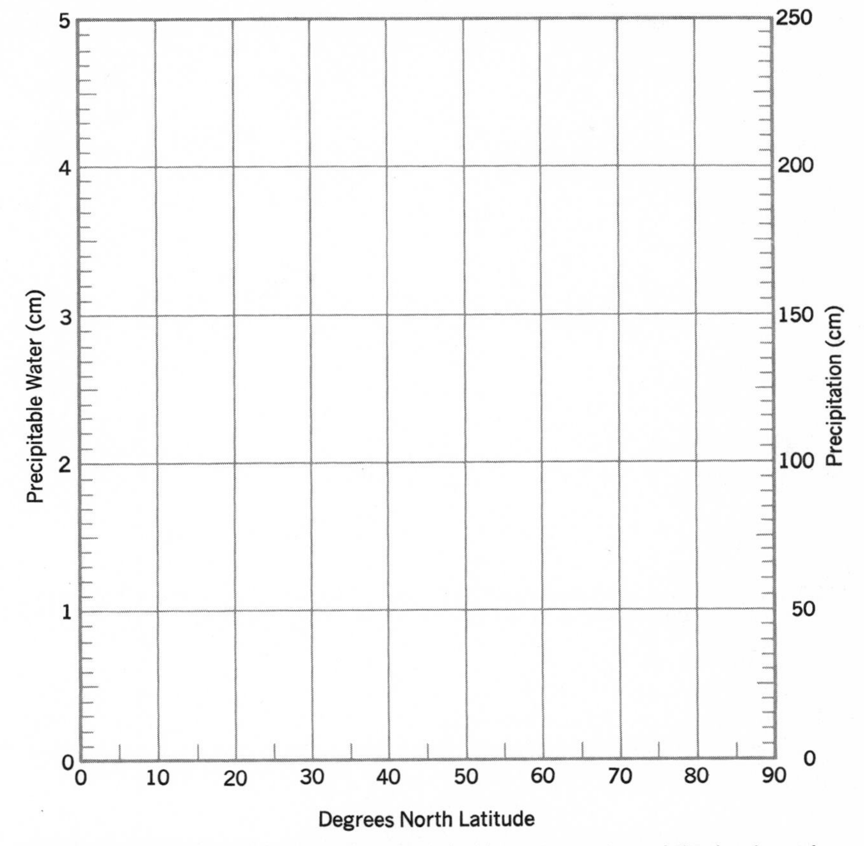

FIGURE 4.3 Relationship between precipitable water and precipitation in various latitude zones.

occur where the amount of precipitable water is relatively large, but the actual rainfall process depends on other factors such as upward motion of the air and the continuous feeding of moisture into the rainfall region by converging winds.

In summary, most precipitation occurs at low latitudes near the belts of greatest evaporation, but some of the water vapor is transported poleward where it is condensed out in the colder air to produce a great amount of cloudiness without yielding much precipitation.

EXERCISE 15

TABLE 4.2
Mean Annual Precipitation and Precipitable Water by Latitude Zones (After W. D. Sellers)

LATITUDE DEG. N	0–10	10–20	20–30	30–40	40–50	50–60	60–70	70–80	80–90
PRECIPITATION (cm)	193	115	79	87	91	79	42	19	12
PRECIPITABLE WATER (cm)	4.1	3.7	2.6	1.9	1.5	1.2	0.9	0.7	0.5

Plot the above data on the graph of Figure 4.3.

a) Does it snow very much in Greenland?

b) Why then is the huge glacier there?

c) Why is there a secondary minimum of precipitation at latitudes 20–30° and a secondary maximum of precipitation at latitudes 40–50°?

d) Would you expect the precipitable water over the United States to be greater during the warm or the cold seasons?

e) When would you expect most places in the United States to get their maximum precipitation, in summer or in winter?

4.2 FURTHER REMARKS ON EVAPORATION

Evaporation rates are controlled in the first instance by the temperature of the evaporating surface; in the special case of ice and snow, evaporation tends to be substantially smaller than from liquid surfaces. Further, evapotranspiration rates from vegetation are strongly influenced by the plant temperature, although here the availability of soil moisture is also

important. In addition, evaporation rates are affected by the dryness of the air and the speed and character of the wind. When the overlying air is dry, the vapor decreases upward more rapidly; turbulent eddies then replace moist air near the surface with dry air from aloft, thus speeding up the evaporation rate. From experience, there is a strong notion that higher wind speeds increase the rate of evaporation, when it is actually the increased intensity of turbulence that produces this effect. The wind speed, regardless of turbulence, directly affects evaporation only when dry air is being transported over wet surfaces.

The secondary controls exerted upon evaporation by the dryness of the air and the character of the wind can become important in several climatic situations: In the lee of large land masses, evaporation rates tend to be large when dry air flows out over water. A notable example of this occurs just off the east coasts of the continents in the zone of prevailing west winds. Also irrigated vegetation in hot climates suffers extreme evapotranspiration, especially when the wind is strong.

A special situation with respect to evaporation arises over snowfields. It should be remembered that heat is required for both evaporation and melting of snow. However, the amount of heat required to evaporate one gram of snow is about 680 calories, whereas only 80 calories are needed to melt the same amount. Therefore, evaporation from snow packs consumes large quantities of heat, that would otherwise be available for melting. That is to say, warm dry air will not produce the depletion of snow or ice surfaces that warm moist air does. When the air is dry, a relatively small quantity of snow evaporates using up a large amount of heat; on the other hand, when the air is moist so that little evaporation can take place, the same amount of heat will melt much ice. This helps to account for the survival of the glaciers at high latitudes where there is little precipitation, because the surface air there is usually quite dry and the snow evaporates only slowly.

How different factors cooperate to produce specific variations of evaporation is shown in Figure 4.4; the annual amount of water evaporated from lake surfaces at different altitudes in Nevada is greatest at low elevations. The lake surfaces are not only warmer there, but also remain ice-free for a longer period. The decisiveness of these effects becomes clear when it is noted that the air is usually drier and the wind stronger at higher elevations.

EXERCISE 16

If Phoenix, Arizona, proposed to build a large reservoir to contain the spring runoff from the mountains, where would you advise them to build it, in the nearby lowlands or up in the mountains? Give your reasons in terms of potential evaporation losses and other pertinent factors.

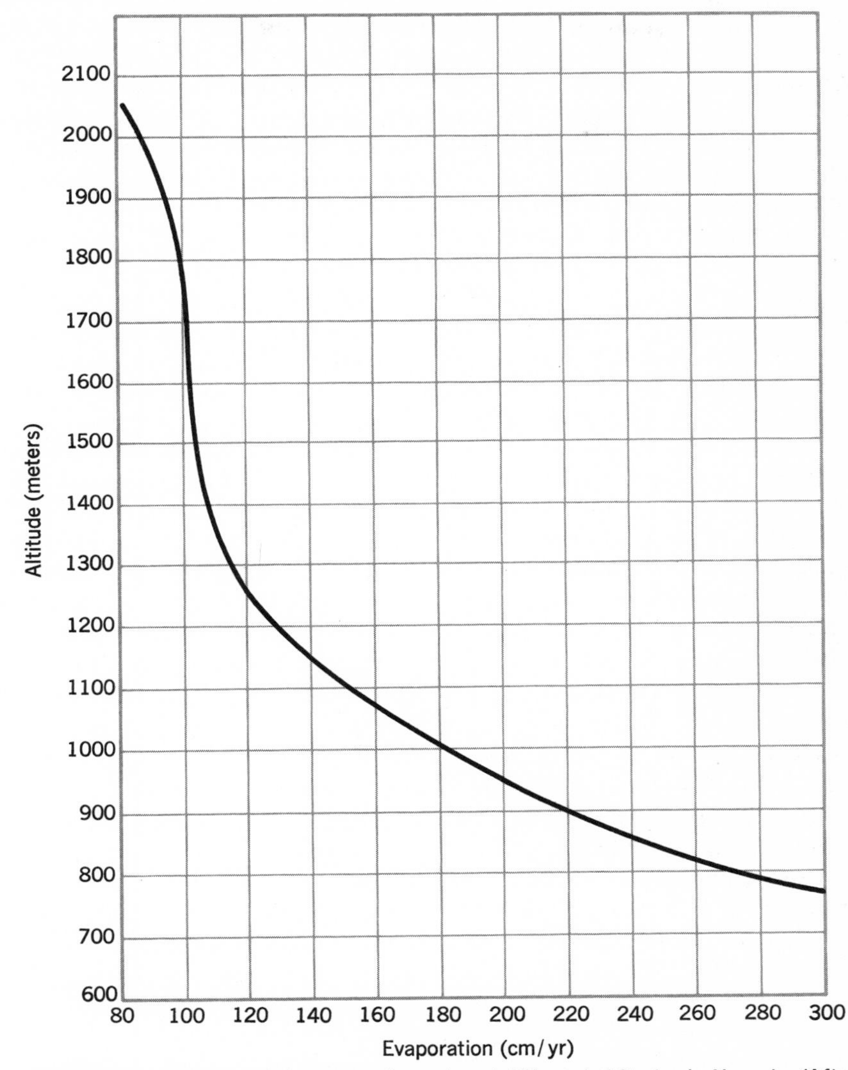

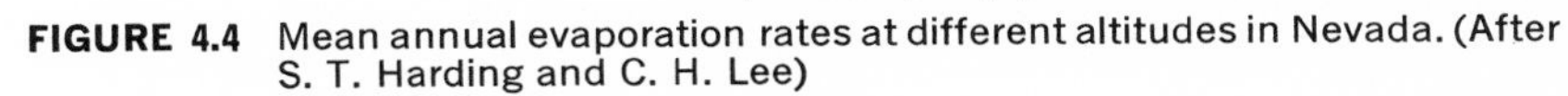
FIGURE 4.4 Mean annual evaporation rates at different altitudes in Nevada. (After S. T. Harding and C. H. Lee)

The heating of the atmosphere in the tropics involves water vapor to a considerable extent. Inasmuch as the precipitation amounts in the ICZ belts are very large, comparably large quantities of latent heat are released into the atmosphere by condensation and are, in effect, left behind by the subsequent precipitation process. More than 80 percent of the atmosphere's net radiative heat loss is made up by the release of this latent heat. As a matter of fact, some authors have called the showers and thunderstorms of the tropics the "Firebox of the Atmosphere." A particular case in which this process truly gets out of hand is the hurricane; much of the energy of its violent winds stems from the latent heat released in heavy squalls. Thus water plays its many roles, moderating otherwise harsh temperature extremes, transporting heat, supporting life, and occasionally taking it.

5 interaction of air, sea, and land

5.1 OCEAN EFFECTS

The atmospheric circulation arising from the large-scale temperature gradient produces different effects at diverse places. What is more, the atmospheric circulation induces circulations in the oceans that lead to modifications of the oceanic surface temperatures. This, in turn, greatly alters the air temperatures at some locations; for example, western Europe, although at relatively high latitudes, is quite warm, whereas the California coast at lower latitudes is relatively cool. These distortions in the temperature patterns of the earth feed back into the atmospheric circulations. Hence, we are faced with a number of interactions between air, sea, and land which may exert decisive controls on climate.

Under otherwise identical conditions, the air temperature rises faster and to a higher degree over land than over water. Under the influence of large bodies of water, air-temperature changes from day to night and from summer to winter are much less than they are at continental locations. Further, temperature changes from one day to the next, called interdiurnal temperature changes, are smaller in maritime than in continental regions. Finally, the large heat storage capacity of water results in a time lag of the maximum temperature not only of water itself, but also of the air over and near the water.

What is true locally is also true for the earth as a whole. Because 39 percent of the northern hemisphere is continental against 19 percent of the southern hemisphere, the average annual temperature range is 14°C for the northern hemisphere as compared to 7°C for the southern hemisphere. Moreover, the higher average summer temperatures on the northern hemisphere cause the earth as a whole to show "seasons:" The average air temperature for the entire earth is 61°F in July, 54°F in January, although the radiation conditions are practically the same throughout the year; if anything, radiation would have to favor January when the earth is slightly nearer the sun than it is in July.

The greater continentality of the northern hemisphere results in a greater annual temperature range at most latitudes. However, it should be recalled that the range of solar elevation during the year also controls the range of temperature, especially so at higher latitudes. Therefore, the temperature range is not a precise measure of continentality unless localities at the same latitudes are compared. Such comparisons reveal the differences in continentality between the two hemispheres, as can be seen in Table 5.1.

TABLE 5.1
Average January and July Temperatures (°C) on Both Hemispheres

	North Latitudes							South Latitudes					
	90°	75°	60°	45°	30°	15°	0°	15°	30°	45°	60°	75°	90°
Jan.	−41	−29	−16	−2	14	24	26	26	22	12	1	−4	−11
July	−1	3	14	21	27	27	26	22	15	6	−10	−30	−42
Temp. Range													

EXERCISE 17

Fill in the temperature ranges in Table 5.1 and plot these values in Figure 5.1.

a) Why is the difference in temperature range between the two hemispheres greatest in the middle latitudes?

b) Why are the temperature ranges large near the poles on both hemispheres?

c) Considering latitudes 45° and 60° both north and south, explain the differences between the two hemispheres regarding the summer temperatures in each; do the same for the winter temperatures.

Just as for the hemispheres, the annual temperature range at individual stations is a good index of continentality or, in an inverse sense, of the extent of maritime climatic influence. As an example, the average monthly and annual temperatures of Eureka, California, and Omaha, Nebraska, are presented in Figure 5.2; these stations are at the same latitude and have the same mean annual temperature of 52°F.

EXERCISE 18

a) Identify the curves with the appropriate station; on what basis do you decide?

b) When does the highest monthly temperature occur at each station? Why is there a difference?

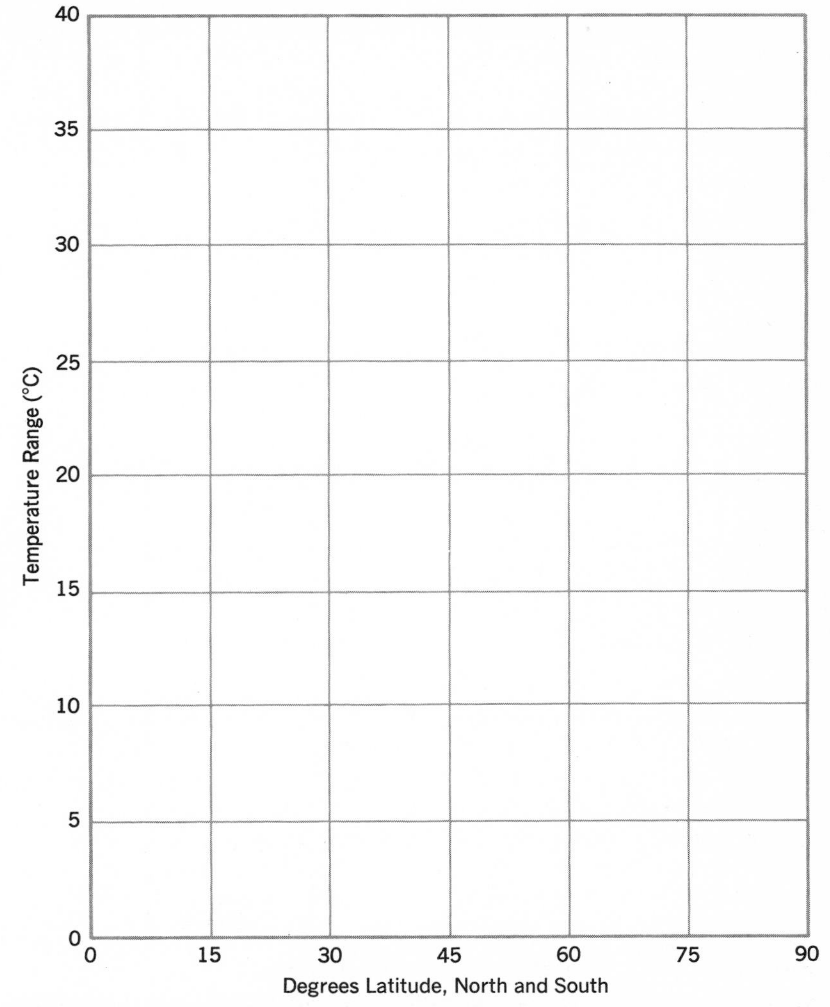

FIGURE 5.1 Mean annual temperature ranges on the northern and southern hemisphere.

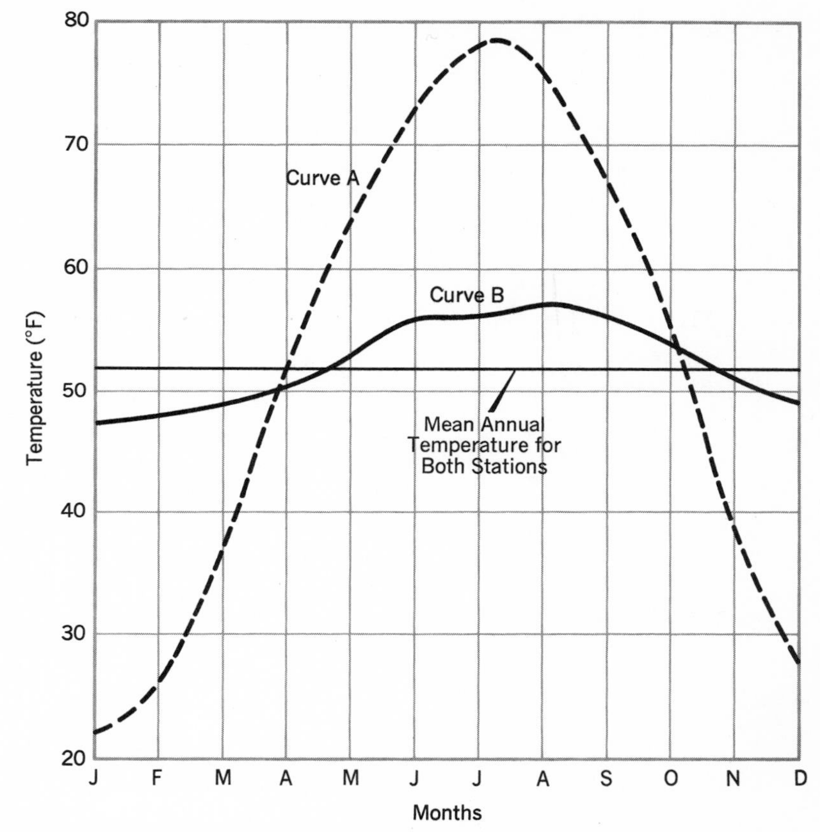

FIGURE 5.2 Mean monthly air temperatures at Eureka, California, and Omaha, Nebraska.

Not only is climate influenced by the presence of oceans in a static sense, but also by the motion of the water as manifest in the essentially wind-driven ocean currents. In Figure 5.3 some of the ocean currents in the Atlantic and Pacific Oceans are depicted by arrows: the average positions of the surface high- and low-pressure systems mentioned in Chapter 3 are also indicated by appropriate labels. The rotation of the winds around these pressure systems is reflected by the ocean currents. These currents are capable of transporting large amounts of heat; one familiar warm current is the Gulf Stream which modifies the climate of northwestern Europe very substantially. The magnitude of this effect is revealed by the data of Table 5.2; the four stations used are located at roughly 63°N and are marked by numbers on the map of Figure 5.3.

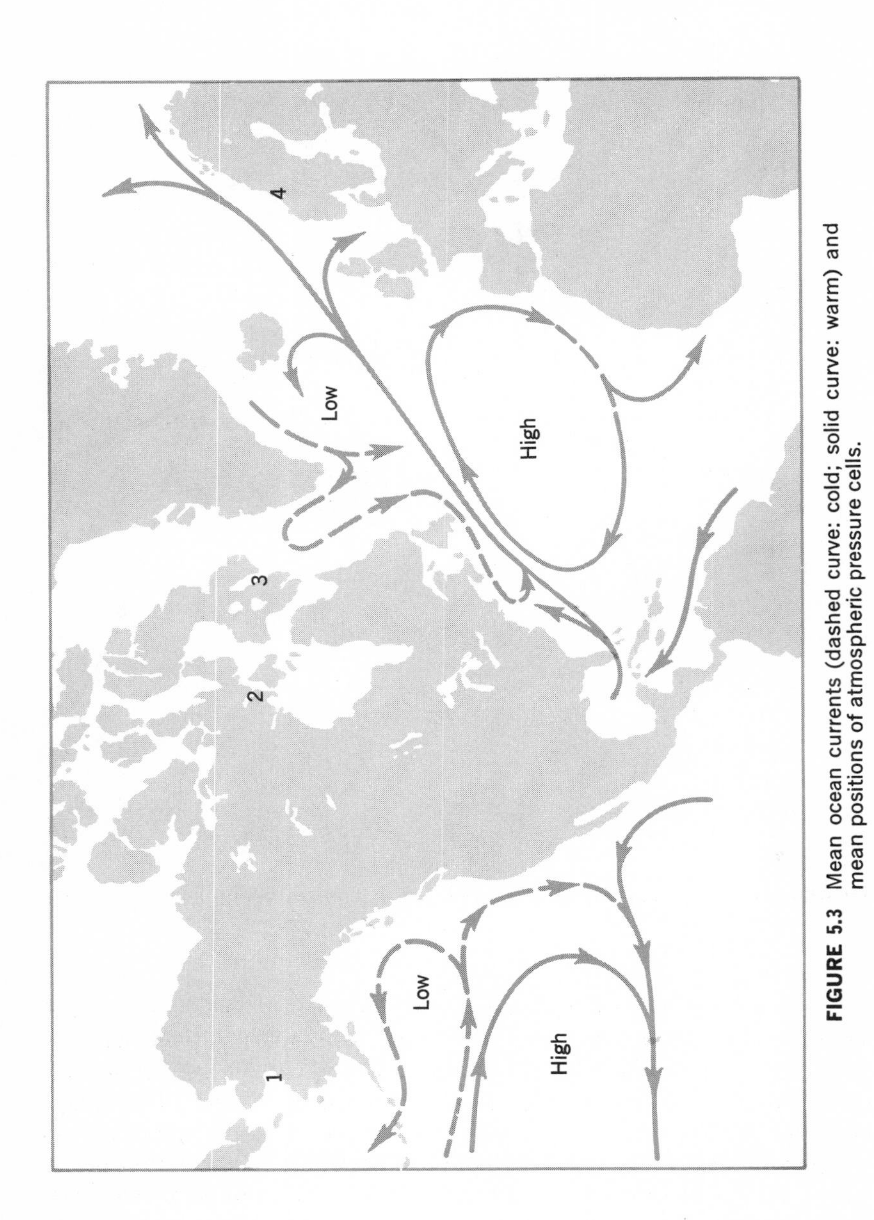

FIGURE 5.3 Mean ocean currents (dashed curve: cold; solid curve: warm) and mean positions of atmospheric pressure cells.

EXERCISE 19

a) Plot the temperatures of Table 5.2 on the diagram of Figure 5.4. Explain the temperature differences and similarities between the North American stations and compare them with temperatures at Trondheim.

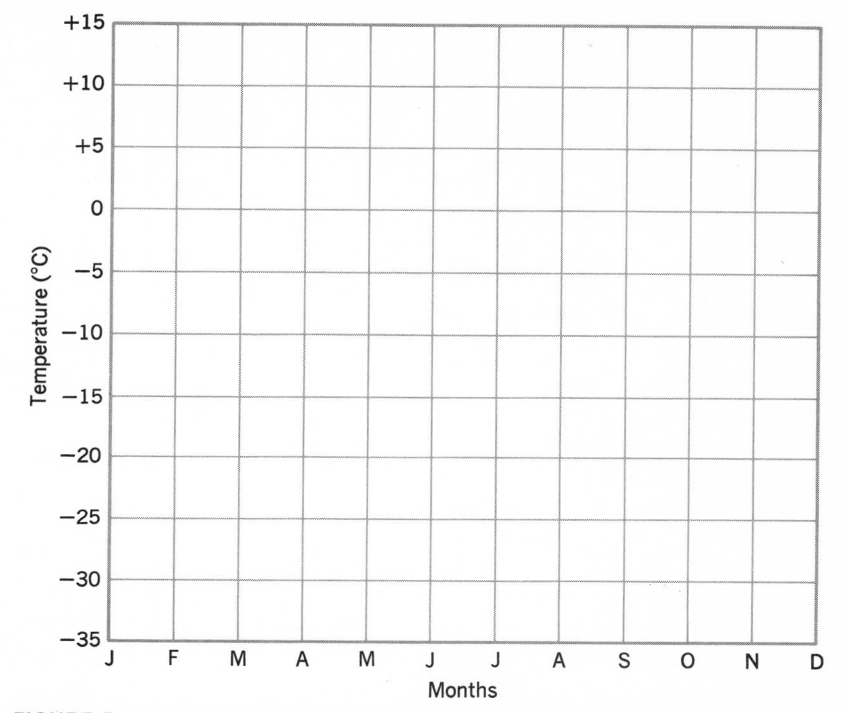

FIGURE 5.4 Annual variation of temperature at four different stations at about 63° north latitude.

b) What causes the differences in annual precipitation between Chesterfield and Trondheim?

c) Fill in the mean annual temperature ranges in the space provided in Table 5.2; how can you show that despite the strong influence of the Gulf Stream, the control by radiant solar energy is dominant?

TABLE 5.2
Annual Variation of Temperature and Total Precipitation at about 63°N Latitude

Month	Holy Cross Alaska 62°10′N;159°45′W #1	Chesterfield Inlet, Canada 63°20′N;90°43′W #2	Frobisher Canada 63°45′N;68°33′W #3	Trondheim Norway 63°25′N;10°27′E #4
Jan.	−18°C	−31°C	−26°C	−4°C
Feb.	−15	−32	−25	−4
Mar.	−11	−24	−21	−1
Apr.	−2	−16	−13	3
May	7	−6	−2	8
June	13	3	4	11
July	14	9	8	14
Aug.	12	9	7	13
Sep.	7	3	3	9
Oct.	−2	−6	−5	5
Nov.	−10	−17	−14	2
Dec.	−18	−26	−22	−2
Year	−2°C	−11°C	−9°C	4°C
Precip. (inch)	16.8	9.7	18.0	33.6
Temp. Range				

A cold ocean current can produce surprisingly large temperature differences on a much smaller scale. An example of this can be found in northern New Jersey where a current of cold water from the northeast hugs the coastline. In Table 5.3 the mean maximum and mean minimum temperatures for New Brunswick and Sandy Hook, New Jersey, are given. The two stations are roughly 20 miles apart as shown in Figure 5.5. Under solar control, the mean annual temperature at both stations is the same, 53°F.

TABLE 5.3
Mean Monthly Maximum and Minimum Temperatures for New Brunswick and Sandy Hook, New Jersey

	Maximum Temperatures (°F)												
	Jan.	Feb.	Mar.	Apr.	May	June	July	Aug.	Sep.	Oct.	Nov.	Dec.	Year
New Brunswick	41	42	50	62	73	81	86	84	77	67	54	42	63
Sandy Hook	38	38	45	56	67	76	82	80	74	64	52	41	60
DIFFERENCES													
	Minimum Temperatures (°F)												
New Brunswick	24	24	31	40	50	59	64	63	56	45	36	26	43
Sandy Hook	27	26	33	42	52	61	67	67	61	51	40	30	46
DIFFERENCES													

EXERCISE 20

Fill in the temperature differences in Table 5.3; be sure to keep track of the signs.

a) Why does the largest difference in maximum temperature occur during spring (this is typical for land-sea daytime temperature contrasts)?

b) Considering the fact that the minimum temperature usually occurs at night, discuss and explain why the largest differences in minimum temperature between the two stations are found in autumn.

The higher minimum temperatures at Sandy Hook are symptomatic of the higher temperatures during the colder seasons, in general. This causes more of the winter precipitation to fall in the form of rain rather than snow, as is evident from the fact that the total mean annual snowfall at New Brunswick is 25 inches, while at Sandy Hook it is only 17 inches, despite the fact that the total precipitation is similar, namely, 46 inches at New Brunswick and 42 inches at Sandy Hook.

EXERCISE 21

Considering the fact that the ratio of snow depth to its rainwater equivalent is usually assumed to be 10:1 on the average, compute the percentage of the total precipitation that falls as snow at the two stations.

In summary, ocean currents help to ameliorate the temperature contrast between the tropics and the polar regions, augmenting the poleward

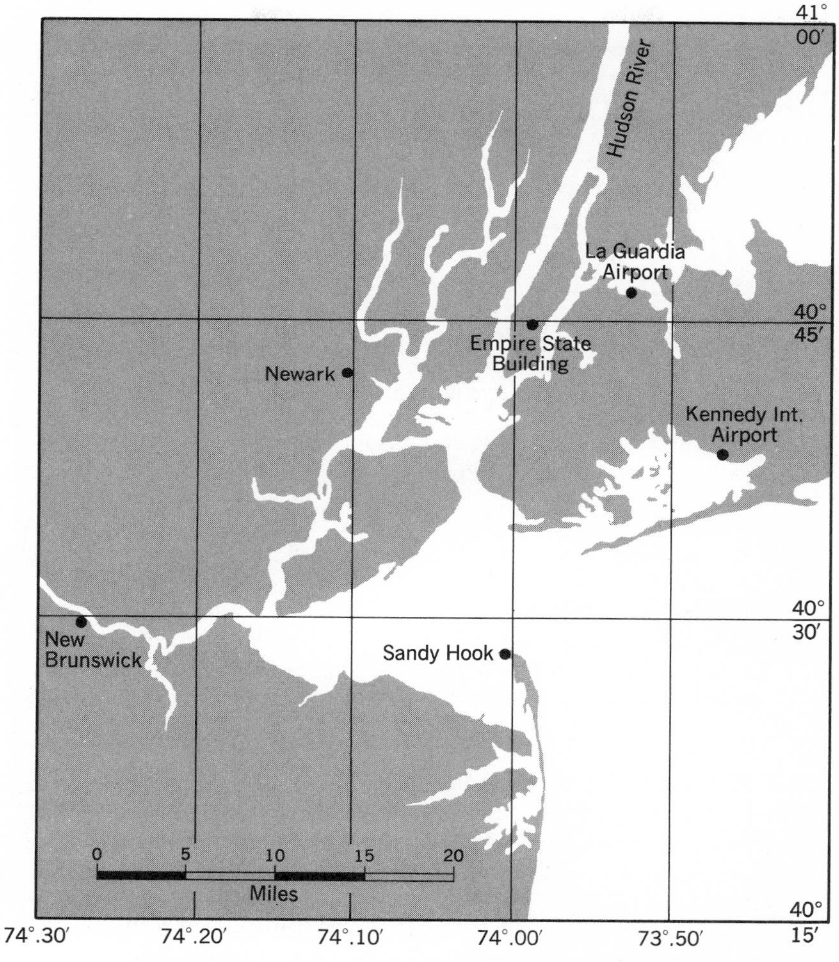

FIGURE 5.5 Location of New Brunswick and Sandy Hook.

heat transport effected by the Hadley cell in low latitudes and by the large vortices of higher latitudes. In addition to direct thermal effects, ocean currents also contribute to other significant climate modifications. One has, for example, only to think of the extensive fogs observed in the Grand-Banks area of Newfoundland and the well-known fogs near San Francisco to recognize the consequences of the interaction between sea and air. In both cases, warm moist air flows over a cold ocean current and is cooled to below its dew point to cause persistent fogs. The specific

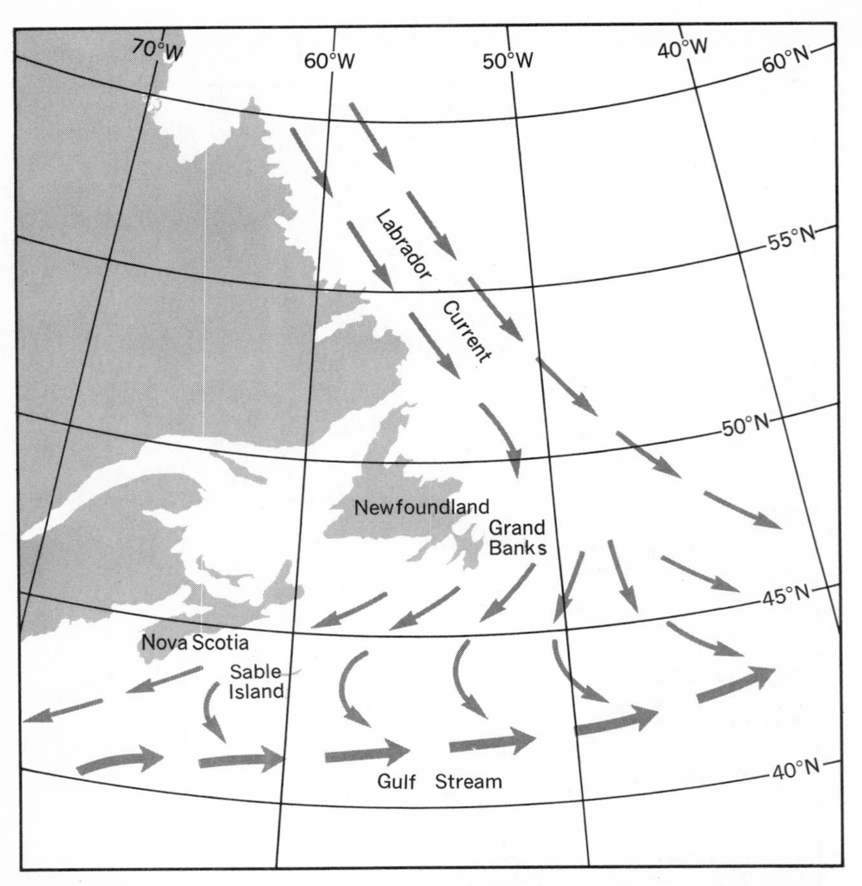

FIGURE 5.6 The Grand Banks area with Labrador Current and Gulf Stream.

situation of the Grand-Banks is shown in the map section of Figure 5.6; here it is evident that any air advected, that is, transported from southerly directions has been warmed and humidified by flowing over the Gulf Stream and is then cooled by the cold Labrador Current.

5.2 MONSOONS

Strictly speaking, a monsoon is defined as a seasonal wind; changes in the predominant wind direction from one season to the next can lead to significant changes in other climatic elements, such as rainfall. And because rainfall markedly changes when the monsoonal windshift occurs, it is quite understandable that one immediately thinks of rainfall when "monsoon" is mentioned. However, cloudiness, water vapor, temperature, and so on, may also be altered in response to the change in wind direction. Monsoons are most prominent in the tropics and subtropics, especially in India, southeast Asia, Australia, and parts of Africa around the Arabian Sea. But some regular wind shift can be observed in many other locations, both inside and outside the tropics. In fact, even in the eastern United States a weak monsoon is evident with west to northwest winds more common in winter, southwest winds in summer.

Monsoon effects are greatly variable from day to day, month to month, and year to year. Even in India terms like "rainy season" can be quite misleading. The rain is usually quite irregular and showery; widespread "breaks" in the monsoon are not infrequent, and much of

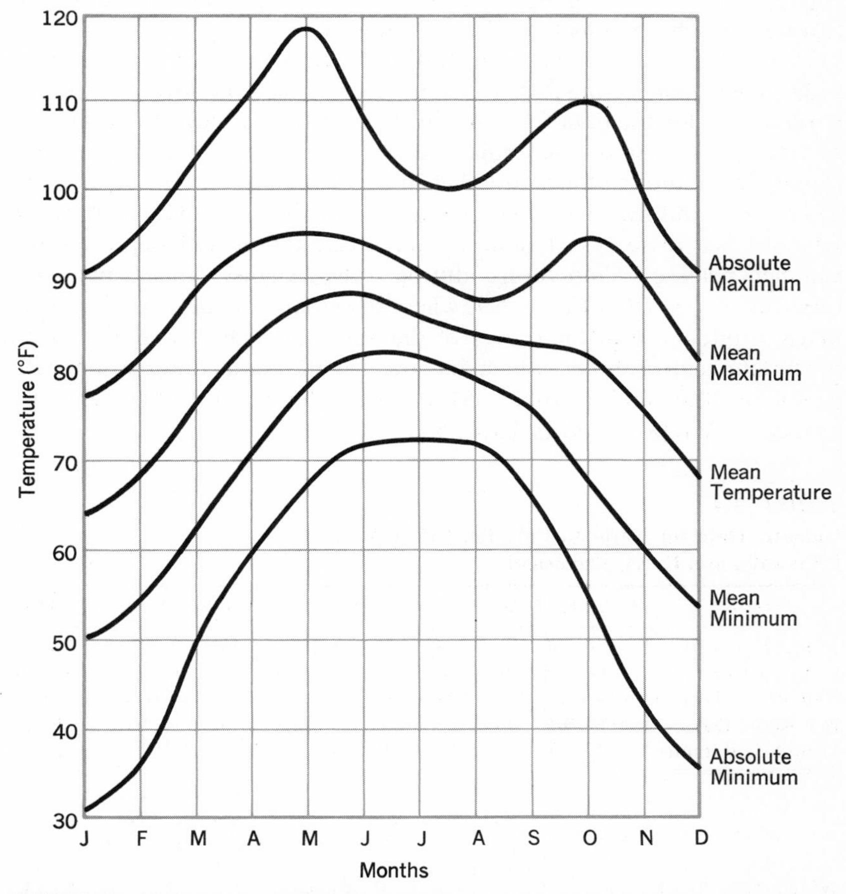

FIGURE 5.7 Various temperature averages for Karachi, Pakistan. (After M. B. Pithawalla and K. M. Shamshad)

the total rainfall may occur on a relatively few days. What is more, local terrain effects can produce a rainy season in one part of a country and a dry season in another part at the same time as, for example, in Viet Nam.

Historically, monsoons were thought to be a simple manifestation of the response of the winds to seasonal heating and cooling differences between continents and oceans. But then, heating and cooling are under solar control and therefore rather regular, so that the bewildering irregularity of the observed monsoons became puzzling. More recently, attention was focused on the less regular seasonal shift in the Hadley-cell circulation, especially in the upper atmosphere, as the primary cause of monsoons. Most modern theorists hold that the two effects are related to each other in many complex ways and that they act in concert with the topography to generate the observed climatological features. Much remains to be learned, however.

Here, then, is an excellent illustration of land, sea, and air acting on each other, and being acted upon, to produce important climatic variations. Climatic data for Karachi, West Pakistan, near 25°N latitude on the west coast of the Indian subcontinent, show these variations rather well, although the station lies in a region that is virtually a desert. There, the wind shift with the attendant advection of moist air from the Arabian Sea increases cloudiness and precipitation and depresses the summer peak of temperature during July, August, and September; this can be seen from Figure 5.7, where the summer depression is particularly pronounced in the curves of the absolute highest and the mean maximum temperatures, which have a double annual oscillation with peaks in May and October. Monsoonal changes of other climatic elements are reproduced in Table 5.4.

TABLE 5.4
Climatic Data for Karachi, Pakistan (After M. B. Pithawalla and K. M. Shamshad)

	JAN.	FEB.	MAR.	APR.	MAY	JUNE	JULY	AUG.	SEP.	OCT.	NOV.	DEC.
Cloudiness (1/10)	3	4	3	3	3	6	8	9	7	2	3	3
Sunshine (hours)	9	9	10	10	10	8	4	4	8	9	9	9
Rainfall (inches)	0.19	0.23	0.03	0.27	0.16	0.72	6.16	0.77	0.38	0.00	0.14	0.12
No. Rainy Days	0.5	0.5	0.1	0.4	0.3	0.9	4.7	1.0	0.9	0.0	0.3	0.3
Wind Speed (mph)	7	7	8	10	13	14	14	14	12	6	7	6

	PERCENT FREQUENCY OF WIND DIRECTIONS								
	N	NE	E	SE	S	SW	W	NW	Calm
Winter (Oct./May)	23	22	3	0	0	5	21	8	18
Summer (June/Sep.)	0	1	2	0	0	25	60	4	8

EXERCISE 22

a) What percentage of the total annual precipitation falls during the summer monsoon?

b) What is the average amount of rain per rainy day during the winter monsoon? During the summer monsoon?

c) Considering wind direction and speed, is the summer or the winter monsoon better developed?

d) Which of the climatic elements indicate the duration of the summer monsoon most significantly?

e) In describing the climate of Karachi in July, it would not be correct to say "rainy." How should it be described?

It was noted in Chapter 3 that the tropical Hadley cell tended to migrate northward and southward with, but lagging somewhat behind, the sun. This behavior is much more pronounced in the northern than in the southern hemisphere, suggesting that the greater continentality of the northern hemisphere really does contribute to the wind-pattern changes associated with this migration. Further, the northward progression of the northern ICZ is greatest over the great land mass of Asia, reaching the southern slopes of the Himalayas between 25° and 30° north latitudes and even farther north along the China coast. The seasonal shift of the weaker southern-hemisphere ICZ amounts to only a few degrees latitude.

There is no doubt that the poleward oscillation of the ICZ at the surface and the associated upper-air features of the Hadley cells of each hemisphere are responses to solar heating (refer to Figure 3.5). What is not yet clear is the relative importance of conditions in the upper atmosphere as compared to surface heating in causing the monsoon. Surface pressures do diminish considerably over the continents in summer as the air is heated, and they increase in winter as the air cools. The resultant surface pressure patterns are consistent with the observed winds, that is, when the pressure is high in winter over China, the surface winds are easterly over India and Pakistan in accordance with the clockwise rotation of the air about the high-pressure center in the northern hemisphere. Similarly, when the thermal low (caused by the warm air) establishes itself over the continent in summer, westerly winds prevail over the subcontinent, consistent with counterclockwise flow around the low.

However, the great wall of the Himalayas extending to 4–9 km above sea level intervenes. A band of strong west winds aloft, the jet stream, is very persistent over the Himalayas in winter. This jet stream is part of the middle-latitude circulation and related to the cold air over the interior of Asia. As the cold air begins to warm up, the jet stream breaks down and permits the subtropical high-pressure belt aloft to migrate quite suddenly northward. Then a surface low-pressure center develops over West Pakistan and China, and the "burst" of the westerly monsoon occurs over the Indo-Pakistan subcontinent, usually between the middle of May and early June. North of the Himalayas, in eastern China, the summer monsoon is weak, whereas the winter monsoon, which is associated with the cold air and high pressure north of the Himalayas, is very strong. Indeed, northwesterly winds sometimes blow at speeds of more than 50 mph for several days in the area of Goto and Amakusa Islands off the southwest coast of Japan. At the same time, the northeast winter monsoon south of the Himalayas is weak and little more than a resumption of the normal Trade winds.

It must be stressed, however, that these particular monsoon effects vary greatly from time to time and from place to place. Monsoon rainfall consists almost entirely of showers and thunderstorms, which, by their very nature, are sporadic and irregular. Furthermore, there is a substantial amount of rainfall generated by relatively infrequent cyclonic wind circulations that usually migrate westward in the zone of the ICZ. The association of these tropical eddies with the ICZ implies that the monsoon rainfall maximum should appear later in the summer at increasing latitudes. This is borne out by observations; however, local effects, especially orographic ones, can completely obscure the situation as shown by mean monthly rainfall and temperature data from four stations in India and Ceylon in Figure 5.8.

The rainfall peaks caused by the monsoon occurs in May at Colombo, in July at Mangalore, and in August at Jodpur, that is, later at higher latitudes, as the ICZ moves northward. However, there is a second rainfall peak at Colombo, which is, in part, produced by the southward return of the ICZ, but even more by the tropical storms over the Bay of Bengal which are most frequent toward the end of the year. It is due to these storms that the monsoonal rainfall at Madras is obscured by the rainfall peak in November; this is also true for other stations along the east coast of India. Comparing precipitation totals at Mangalore (348 cm = 137 inches) and at Madras (123 cm = 48 inches) at the same latitude but on the east coast, we find the effect of the intervening mountains in the depletion of the moisture supply of the air.

As regards the temperature curves, the reduction of the summer temperature by cloudiness and precipitation, already noted in the Karachi data, is also found at Indian stations causing the warmest weather to occur in spring before the rains come.

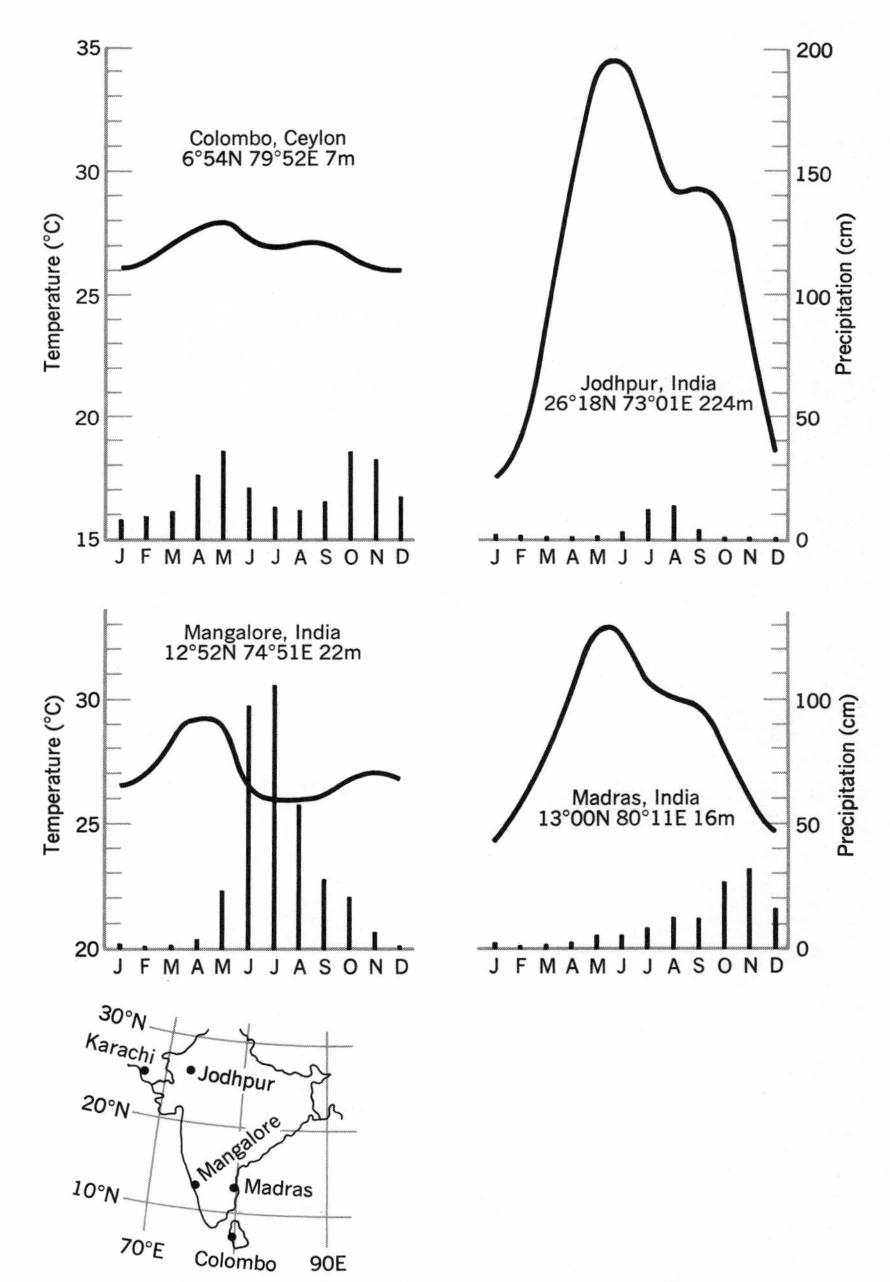

FIGURE 5.8 Mean monthly temperatures and precipitation at four stations in the Indian monsoon region.

EXERCISE 23

a) To which part of India is the term "burst of the monsoon" most appropriate?

b) At Madras, the temperature maximum in May does not appear to be related to any heavy summer precipitation. Is the temperature depression of June, July, and August related to the monsoon? If so, how?

c) Why does Jodpur have the greatest annual temperature range and Colombo the smallest?

5.3 THE SEA BREEZE

The monsoon has an equivalent on a shorter time scale, namely the sea breeze, or where appropriate the lake breeze. At some locations, daytime heating of land surfaces, especially during the warm seasons, either strengthens the onshore flow, weakens the offshore flow, or produces onshore flow that would not otherwise obtain. See Figure 5.9. At night, a reversal occurs, but land breezes are weaker, especially near

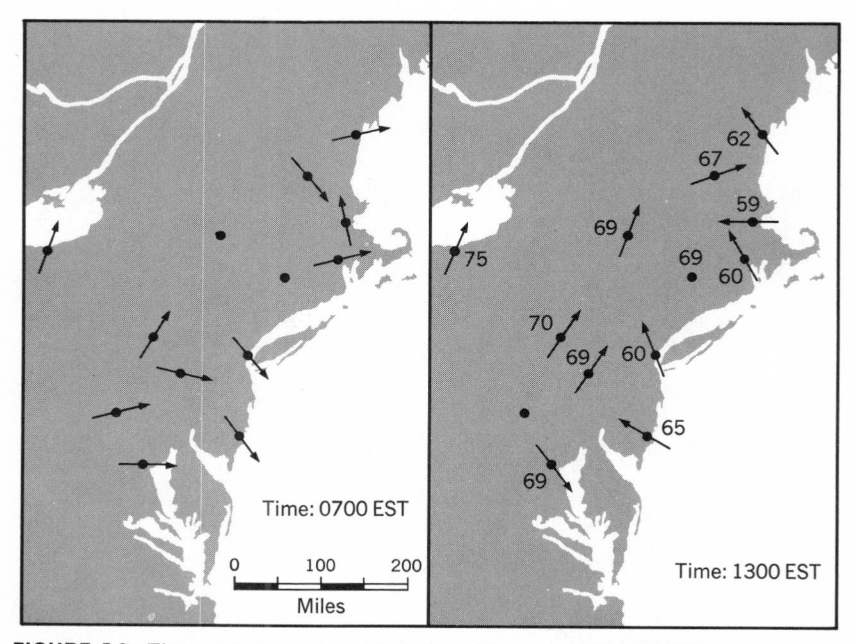

FIGURE 5.9 The sea breeze along the U.S. east coast on 12 April 1968 and its effect on maximum temperatures indicated in °F on the right side of the figure.

the shore, and their climatic effects over land are usually slight. While particular effects of the sea breeze vary from place to place, a few generalizations can be made. On cloudy days or on days when other circulation features predominate, the sea breeze may be entirely absent. Outside the tropics it is strongest in late spring and early summer, when the temperature contrast between land and ocean is greatest. At locations where the "normal" air flow is offshore, the sea breeze raises the day-time relative humidity and cloudiness by its vapor transport and cooling. Finally, at any affected location, it can produce a zone of convergence

FIGURE 5.10 View of Florida Peninsula from the North. Convective clouds cover the Island except over the cooler lakes over which subsidence takes place. Scattered thunderstorms occur over the east coast and several miles off shore. (Courtesy NASA; photograph was taken on one of the manned space flights.)

in the low-level winds along the coast which can manifest itself as a local maximum in precipitation, usually caused by shower and thunderstorm activity. This can be gathered from the satellite photo, Figure 5.10, of the Florida Peninsula in which the convective activity induced by the sea breeze leads to thunderstorms along the coast.

An example of a complete reversal of the wind by the sea breeze with its attendant disruption of the normal diurnal temperature and relative humidity variation is given in Figure 5.11 for Boston, Massachusetts, on 21 May 1964, a cloud-free day. Considering the fact that at Boston the sea breeze is indicated by winds from directions between east and southeast, the onset of the sea breeze can be detected between 9 and 10 o'clock in the forenoon; the wind shift is heralded by the slowing down of the northerly winds around 9 a.m. Simultaneously, the relative humidity begins to rise, and the normal temperature rise is interrupted by the advection of cool air from the sea.

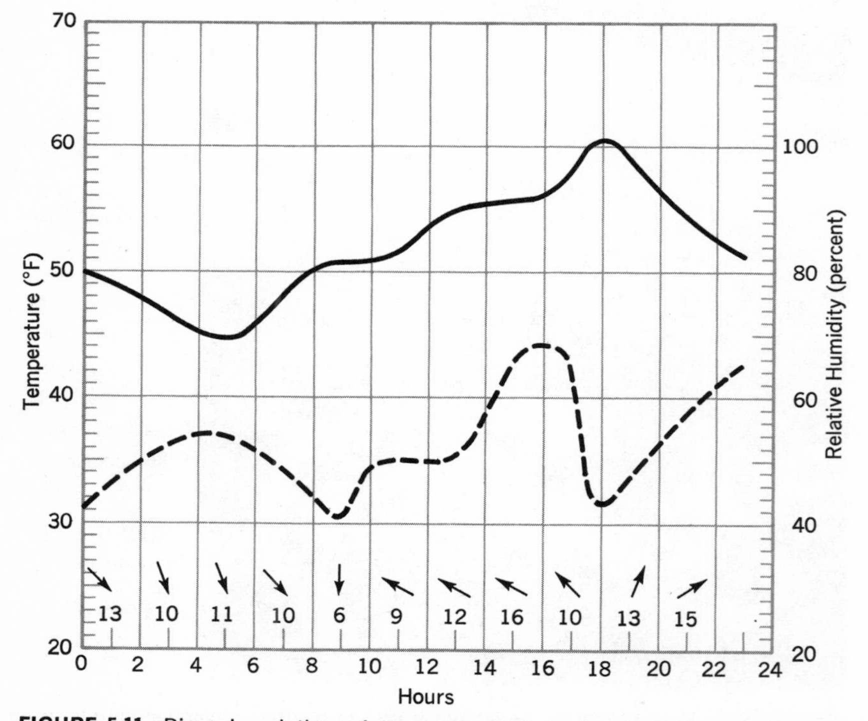

FIGURE 5.11 Diurnal variation of temperature (solid curve), relative humidity (dashed curve), and wind at Boston on 21 May 1964. The arrows fly with the wind; speeds below arrows are given in knots (nautical miles per hour).

EXERCISE 24

a) Assuming a smooth temperature rise continuing the temperature trend established between 6 and 8 a.m. in Figure 5.11, estimate what the maximum temperature would have been at about 15 hours, if there had been no sea breeze.

b) Approximately at what time did the sea breeze end?

c) What clues did you use to determine this time?

In cases of wind reversal after the occurrence of the normal maximum temperature, that is, later in the afternoon, dramatic cooling works its way inland for 10 miles or so. This is illustrated by the diurnal temperature variations recorded in Chicago on 11 July 1957 at the Weather Bureau City station at the Lake shore and at Midway Airport. For comparison, the data for Joliet Airport, some 32 miles inland, are also given (see map of Figure 5.12) in Table 5.5, as are the winds observed at Midway and Joliet.

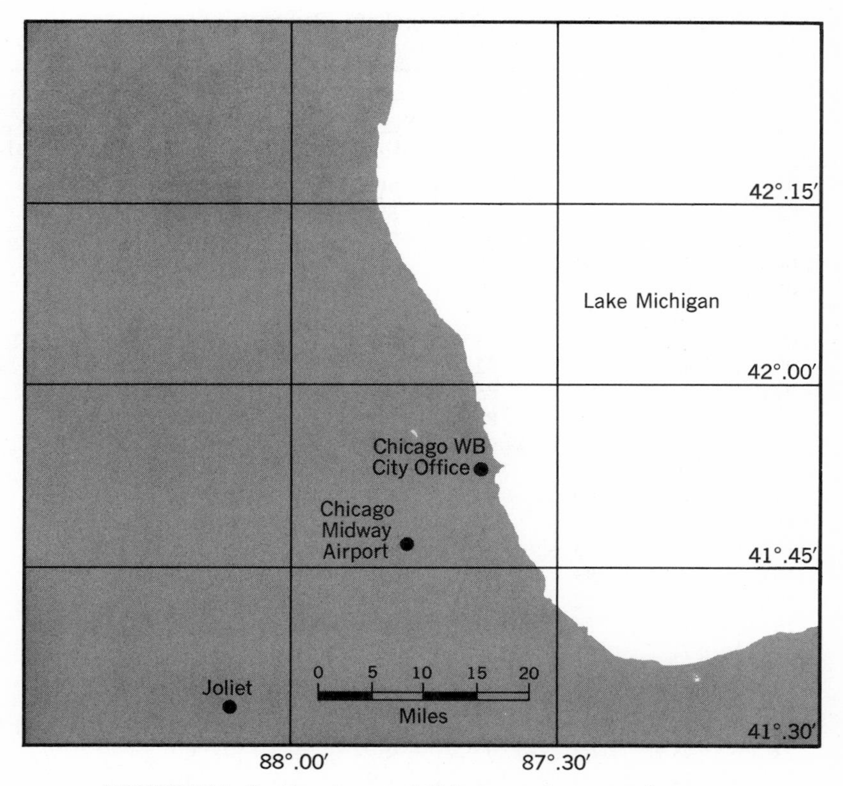

FIGURE 5.12 Sectional map of Chicago and surrounding area.

EXERCISE 25

Plot the temperature data of Table 5.5 on the graph of Figure 5.13.

a) At what time does the Lake breeze set in at the Weather Bureau City Station? At Midway? At Joliet?

b) How is the arrival of the Lake breeze at Midway noticeable in the winds?

c) What is the rate of westward progress of the Lake breeze?

In cities such as Boston, Chicago, and others, people often refer to the sea breeze bringing welcome relief from excessive heat as "nature's air conditioner." In west-coast locations of the tropics, people build simple ducts on top of their houses to catch the cooling sea breezes as shown in Figure 5.14.

TABLE 5.5
Diurnal Temperature Variations Showing Lake Breeze in Chicago Area

Hour	Temperature (°F) City Station	Midway Airport	Joliet	Wind at Midway Dir.	Speed (mph)	Wind at Joliet Dir.	Speed (mph)
01	75	73	67	SW	3	SSW	4
02	73	71	67	SW	3	SSW	5
03	73	70	66	SW	3	SW	2
04	73	70	65	SW	3	SSW	5
05	74	71	66	SW	4	WSW	2
06	76	72	71	SW	5	WSW	4
07	78	75	75	SW	5	WNW	8
08	80	78	82	SW	5	W	8
09	82	82	85	NW	5	NW	11
10	85	86	89	W	6	W	15
11	87	87	90	SW	6	W	15
12	89	89	91	W	7	W	15
13	90	90	92	W	7	NW	15
14	91	92	93	SW	8	W	15
15	91	93	93	W	8	W	17
16	90	93	93	W	7	W	15
17	78	92	92	SW	6	WSW	14
18	72	80	90	E	4	SW	9
19	71	77	84	E	4	SW	5
20	70	75	78	SE	3	N	7
21	74	74	76	S	4	SE	1
22	75	73	73	SW	5	S	13
23	75	74	72	S	4	SW	7
24	74	75	71	SW	3	SW	7

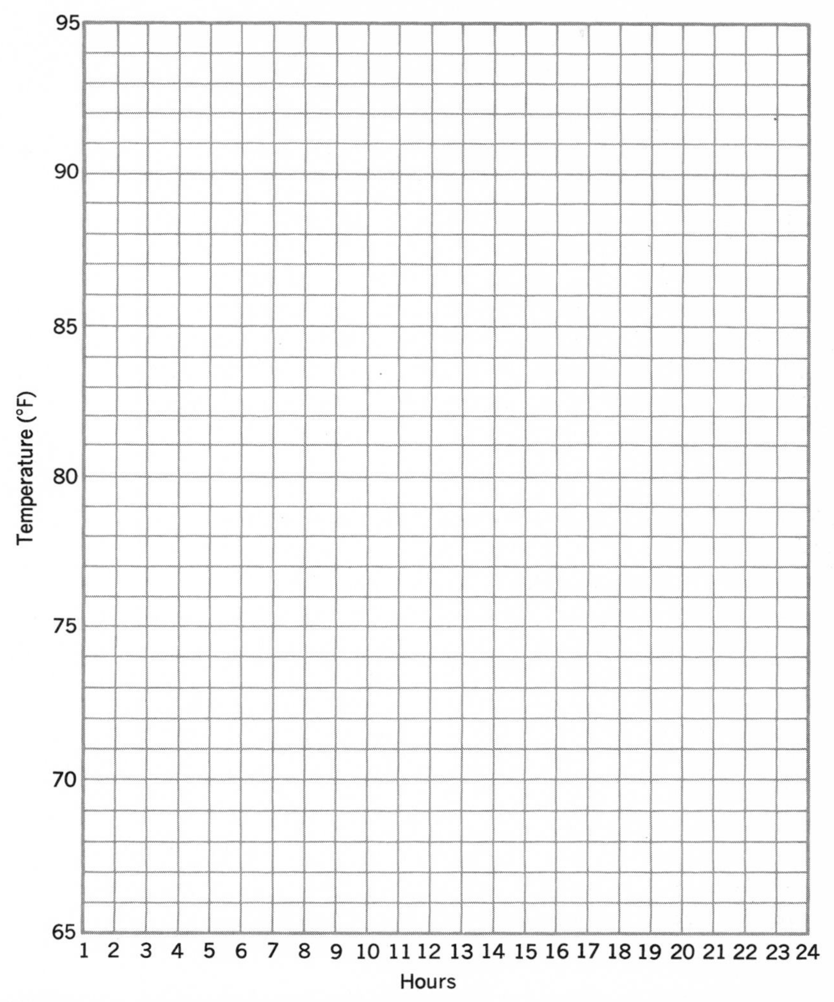

FIGURE 5.13 Diurnal variation of temperatures at the Chicago Weather Bureau Station, at Midway Airport, and at Joliet, illustrating the effect of the lake breeze.

FIGURE 5.14 "Air conditioning" ducts atop a house in Karachi, Pakistan, catching the cooling sea breeze.

5.4 LAKE EFFECTS

Smaller bodies of water influence the climate along their shores in many of the same ways that the oceans do, although to a lesser degree. As an example, data from two stations located on opposite sides of Lake Michigan (for location see Figure 5.15) are presented in Table 5.6; at both stations westerly winds prevail.

EXERCISE 26

a) How is the lake effect revealed by winter temperatures (winter = December, January, February)? The summer temperatures (summer = June, July, August)? The mean annual temperatures?

b) What are the average annual temperature ranges at the two stations, and what do they express regarding the lake effect on continentality?

c) What is the lake effect on precipitation? Which station is likely to have more snow?

Although the effect of Lake Michigan may seem rather subtle, it has a substantial influence on the agriculture of the regions upwind and

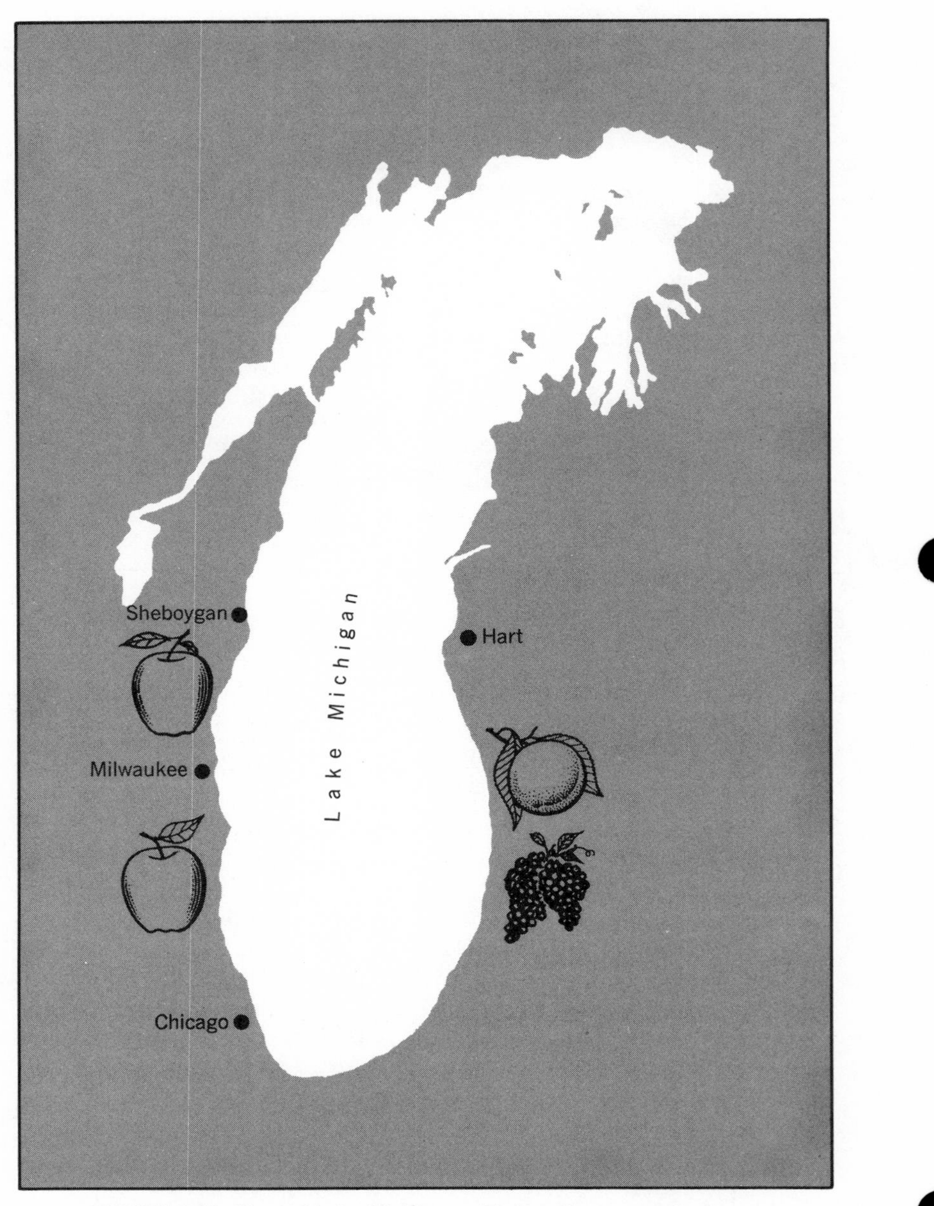

FIGURE 5.15 Map of Lake Michigan with its effect on agriculture.

TABLE 5.6
Mean Monthly Temperatures and Precipitation on West and East Shore of Lake Michigan

Month	Temperatures (°F)		Precipitation (inches)	
	Sheboygan Wisconsin	Hart Michigan	Sheboygan Wisconsin	Hart Michigan
Jan.	20	24	1.20	2.35
Feb.	24	25	1.45	2.00
Mar.	30	31	2.08	2.09
Apr.	44	46	3.18	3.13
May	54	56	3.03	2.86
June	64	65	3.40	3.10
July	70	70	4.07	3.70
Aug.	71	69	2.79	3.12
Sep.	62	61	2.62	2.66
Oct.	51	51	3.06	2.90
Nov.	36	38	2.06	2.94
Dec.	26	28	1.59	2.02
Year	46	47	30.53	32.87
Temp. Range (°F)				

downwind of the Lake; west of the Lake apples are successfully grown, while the moderation of the climate by the Lake will permit the growing of peaches and grapes to the east of the Lake (Figure 5.15). However, there are also effects of lakes that produce rather severe weather: In winter, as long as the water remains unfrozen, there may be sufficient addition of moisture to the air flowing over a lake to permit heavy snowfalls on the downwind shores. This sort of "snow burst" is illustrated by the case of Lake Erie in the storm of 2–3 December 1966, shown in Figure 5.16, in which the curved lines represent lines of equal snowfall. In this particular storm, 52 inches of snow were dumped on Mayville in the southwest corner of New York State. Most of the other affected areas in Ohio, Pennsylvania, and New York observed only two to four inches of snow. Here, the thermal and dynamic effects of the lake surface combined with the topography of the shore regions to create incredible variations in snow depths.

What is true of large lakes is also true of other large water bodies, such as Hudson Bay in Canada. Considering two stations on opposite shores, we see the modification produced not only by the presence of water, but also by its frozen state during a large portion of the year. In Table 5.7, the mean annual temperatures and snowfalls at Churchill

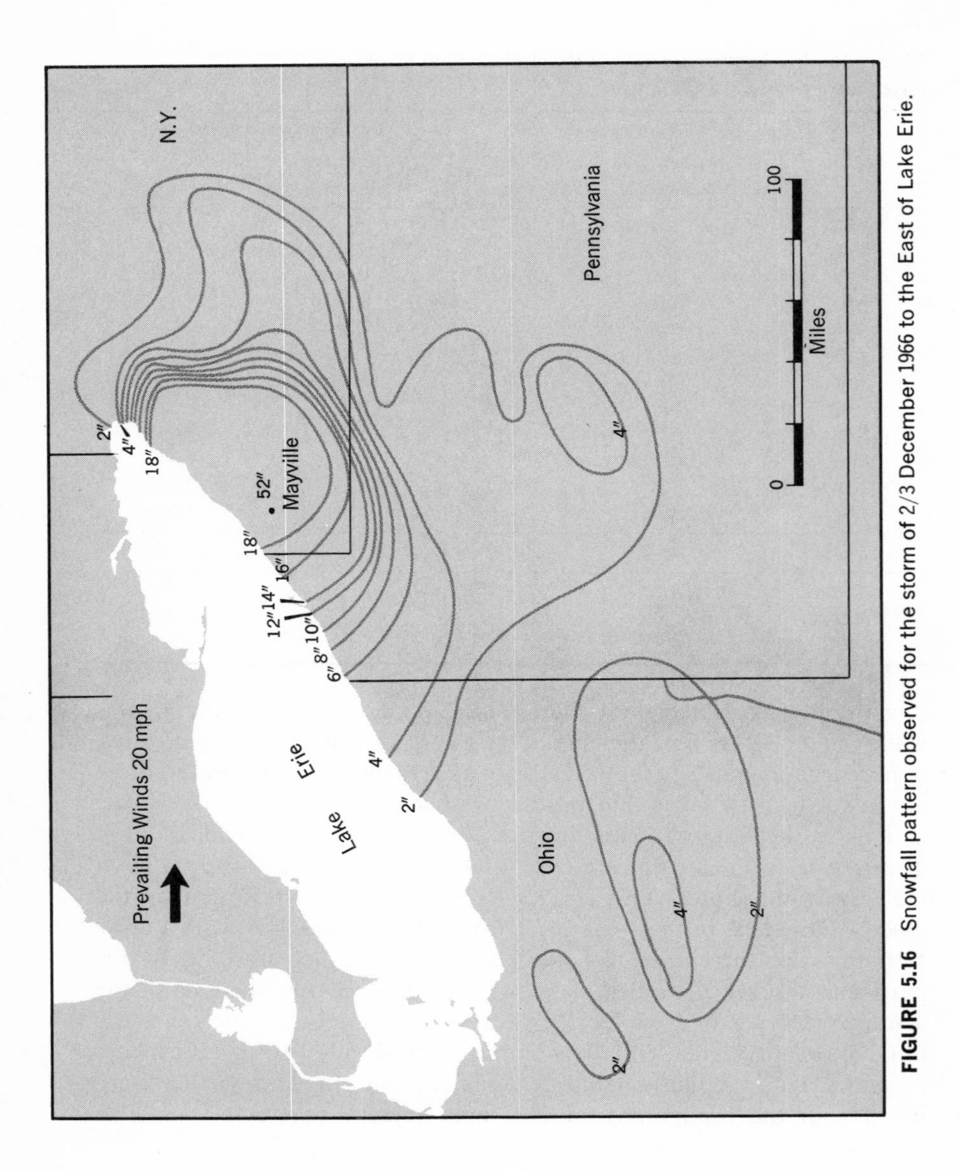

FIGURE 5.16 Snowfall pattern observed for the storm of 2/3 December 1966 to the East of Lake Erie.

(58°45′ N; 94°04′ W; 35 m) and Port Harrison (58°27′ N; 78°08′ W; 20 m) are given; the mean annual precipitation at the two stations is 16.0 and 16.3 inches, respectively. However, the maximum monthly precipitation occurs in August at Churchill, in November at Port Harrison.

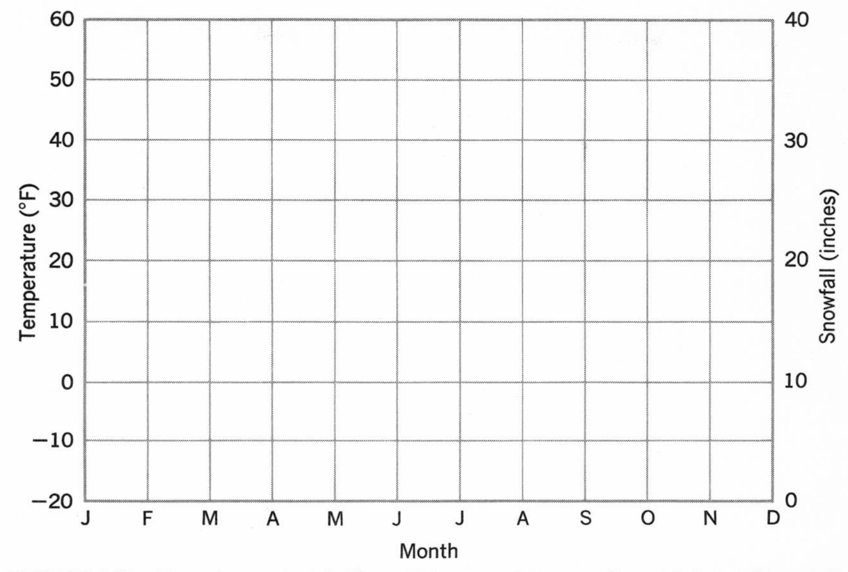

FIGURE 5.17 Mean annual variations of temperatures and snowfalls at Churchill and Port Harrison, Canada.

EXERCISE 27

Plot the temperatures and snowfall of Table 5.7 onto the diagram of Figure 5.17.

a) Considering the fact that both stations lie in the zone of prevailing westerlies, what causes the small shift of the maximum temperature and the delay in cooling at Port Harrison?

b) What percentages of the total precipitation fall as snow at each station?

c) What is the reason for the resulting difference?

TABLE 5.7

Mean Monthly Temperatures and Snowfall at Churchill and Port Harrison

	Temperature (°F)											
Station	Jan.	Feb.	Mar.	Apr.	May	June	July	Aug.	Sep.	Oct.	Nov.	Dec.
Churchill	−19	−17	−6	14	30	43	54	52	42	27	6	−11
Port Harrison	−18	−18	−7	11	29	39	47	47	41	31	17	−2
	Snowfall (inches)											
Churchill	4.8	6.1	8.5	7.7	1.8	1.4	—	—	1.7	8.0	10.3	6.6
Port Harrison	7.1	2.8	7.2	8.4	4.6	2.7	—	—	2.8	15.8	33.0	12.4

6 recurrent and transient phenomena

6.1 SEASONAL SHIFTS OF PERSISTENT CIRCULATION FEATURES IN MIDDLE AND HIGH LATITUDES

It was noted in the discussion of monsoons that in some tropical and subtropical locations, the winds aloft change their pattern at the same time as the surface winds do. The unsteady but inexorable trend of the low-latitude Hadley cells moving northward and southward with the sun was seen to set the stage for large-scale inflow into the continents in the summer hemisphere. The wet season could begin, especially in the region of the ICZ.

Outside the tropics, similar seasonal migrations and changes in intensity of persistent atmospheric flow patterns, pressure systems, and centers of weather action are also observed, and their climatic effects can be quite drastic. Many of these seasonally varying winds could quite properly be called monsoons. But the small upward or downward motions, that are characteristic of the circulation features, are relatively more decisive in determining the rainfall and cloudiness patterns in higher latitudes than they are in the tropics. There, the large-scale upward motions merely provide more or less favorable environments for showers to develop.

It is worthwhile to recall here that upward motion cools the air by expansion, generating clouds, whereas downward motion warms it and suppresses cloud formation. Even though the vertical speeds are very small (of the order of 100 feet per hour), the associated temperature changes are rather large; while showers are common enough outside the tropics, surface heating is, relatively speaking, less important in their formation than are the vertical motions associated with the large-scale circulation features. For these reasons, we examine the seasonal climatic

variations in extratropical regions from the point of view of large-scale lifting and sinking of the air, rather than as monsoons.

A few general remarks hold for all regions outside the tropics: The poleward variation of incoming solar energy is much greater in the winter hemisphere; consequently, the temperature contrast across middle latitudes is greater in winter and so is the strength of the large-scale cyclones and anticyclones that develop in the zone of contrast. These huge and vigorous vortices then transport large amounts of energy poleward and drive the strong westerlies which exist in the upper troposphere in winter. The belt of strong westerlies, the jet stream, is found above the zone of greatest temperature decrease poleward, the so-called polar front.

It follows then that the strength of the interrelated polar front, the jet stream, and the migratory cyclones and anticyclones should be greatest in winter and least in summer, all as a result of the large summer-winter contrast in solar energy received at high latitudes. What is more important to the present discussion, however, is the additional conclusion that the polar front and its associated jet stream will shift poleward in summer as it weakens. This process need not be very steady; in fact, the locations of fronts and jets vary widely from day to day. Nevertheless, the proposition holds: During spring the polar front with its jet moves poleward and weakens; in autumn it tends equatorward and strengthens. In Figure 6.1 the average positions and speeds of the jet stream in summer and winter on the northern hemisphere are shown.

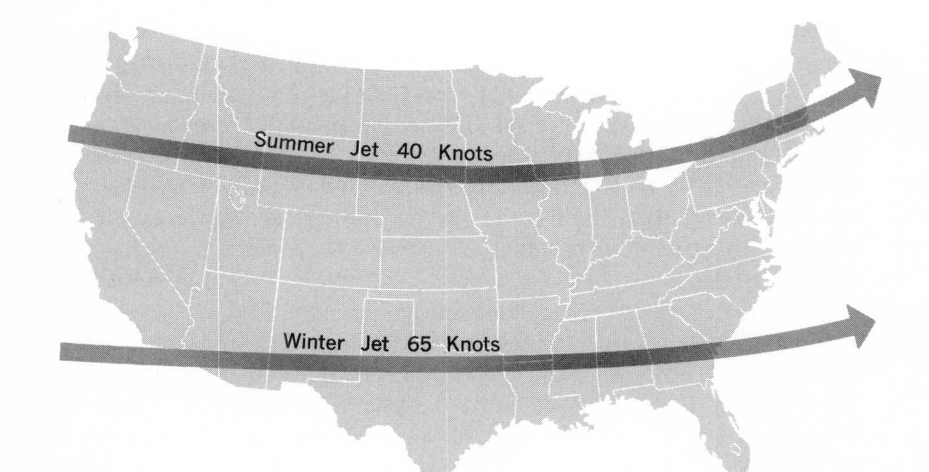

FIGURE 6.1 Average summer and winter positions of the Jet Stream on the northern hemisphere.

On the northern hemisphere, the jet's weakening permits the northward migration of the rather wide belt of subsidence, or sinking air, at the poleward extremes of the tropical Hadley cells. Large changes from winter to summer occur at many locations between about 25 and 40 degrees north latitude. The effect is much smaller in the southern hemisphere, where the Hadley cell migrates but little. During summer, subsidence extends poleward giving clear and practically rainless days to the regions affected, though not in continuous belts around the earth. One such region is found around the Mediterranean Sea, and climates having dry summers and wet winters as a result of the shifts of the subsidence belts are said to be "Mediterranean" or California" climates.

This type of climate is well illustrated by the seasonal shift of the high-pressure cell which is associated with the subsidence off the U. S. west coast. (See Figure 6.2.) As the Pacific High shifts northward in

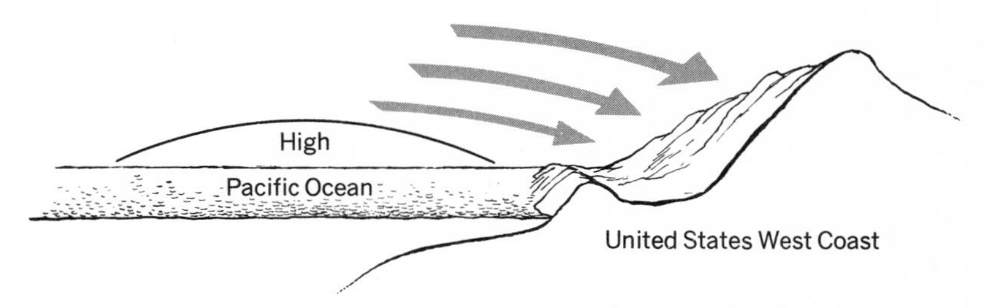

FIGURE 6.2 Subsidence over United States West Coast.

summer, the surface winds blow essentially parallel to the coast line from northern Washington to southern California. Aloft, the dome of high pressure over the Pacific, shown in profile in Figure 6.2, favors strong descent of air, thereby suppressing all but the shallowest of clouds and greatly inhibiting precipitation. In winter, when the polar jet and Pacific High shift southward, cyclones can penetrate into the region of Washington, Oregon, and northern California. Furthermore, the winds around the High now have an onshore component and are lifted by the mountains along the coast (Figure 6.3).

In Table 6.1 the average annual variation of precipitation is given for Portland, Oregon, San Francisco and San Diego, California; for comparison, the data for Atlantic City, New Jersey, are added.

EXERCISE 28

a) Why does San Diego have less precipitation than the other two west-coast stations?

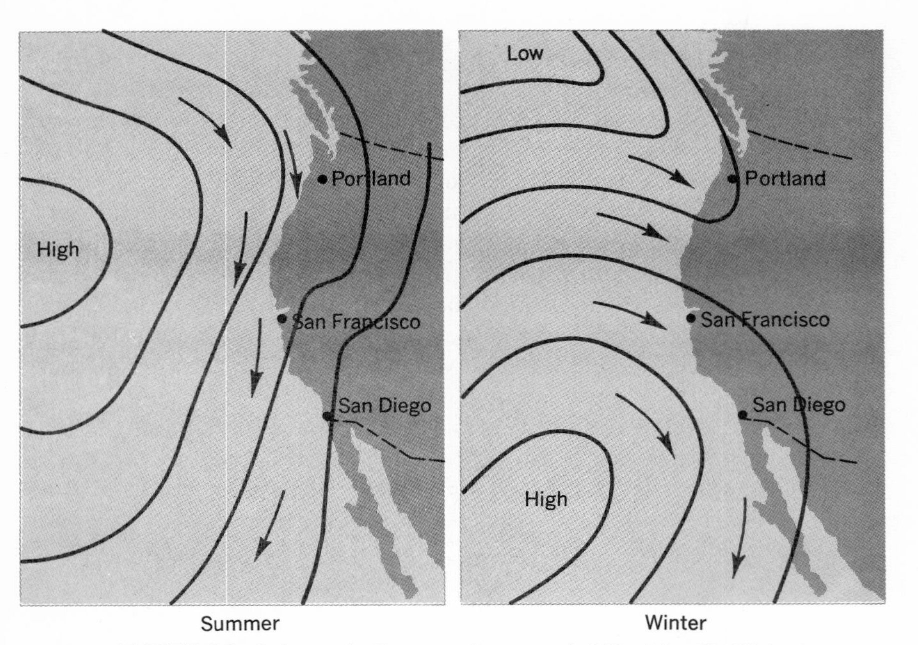

FIGURE 6.3 Schematic diagram of seasonal shift of Pacific High.

b) Why does Portland have more precipitation in summer than the other two stations?

c) Explain the two maxima at Atlantic City in spring and autumn in terms of the Polar Front.

TABLE 6.1
Average Monthly and Annual Rainfall (inches) along U. S. West and East Coasts

Month	Portland	San Francisco	San Diego	Atlantic City
Jan.	6.8	4.7	2.3	3.0
Feb.	4.1	3.3	1.2	3.5
Mar.	3.9	2.4	1.6	4.4
Apr.	2.2	1.6	1.0	3.4
May	2.1	0.6	0.2	3.1
June	1.8	0.1	0.1	2.9
July	0.3	0.0	0.0	4.7
Aug.	0.8	0.1	0.1	5.4
Sep.	1.5	0.4	0.1	2.3
Oct.	3.7	0.5	0.4	3.4
Nov.	4.8	1.6	0.7	4.2
Dec.	5.5	4.3	1.0	3.8
Year	37.5	19.6	8.7	44.1

Elsewhere, the strengthening and northward migration of the so-called Bermuda High in the Atlantic in summer increases the flow of warm moist air from the Atlantic and the Gulf of Mexico into the eastern United States. This raises the precipitation amounts in a monsoon-like fashion and makes the southeastern part of the country just short of unbearably humid at times.

In the south-central provinces of Canada, northward migration of the Polar Front and its jet produces a summer maximum of precipitation. This is also true of regions at similar latitudes in Asia where under the weak summer jet stream cyclones promote shower activity. Because precipitation there is so dependent upon these erratic circulation features, the variability of summer rains is very large, as is typical of all arid and semiarid climates. The recent "Virgin Lands" experiments in the Soviet Republic of Kazakstan in central Asia have demonstrated this rain variability rather graphically by producing large wheat crops in some years, very small crops in others.

The great variability of rainfall in that area is evident from nine years of records at Turgay (49°38′ N; 63°30′ E; 123 m); the highest, lowest, and average monthly rainfalls as well as the annual totals between the years 1952 and 1960 are given in Table 6.2. Note that the average annual total rainfall is less than 8 inches.

TABLE 6.2
Rainfall Records at Turgay, U.S.S.R., 1952–1960

MONTH	MAXIMUM (mm)	MINIMUM (mm)	DIFFERENCE (mm)	NINE-YEAR AVERAGE (mm)	RELATIVE RANGE (percent)
Jan.	23	1		10	
Feb.	37	1		17	
Mar.	22	1		11	
Apr.	47	1		17	
May	39	0		11	
June	52	0		18	
July	116	3		36	
Aug.	58	0		18	
Sep.	28	0		11	
Oct.	39	0		21	
Nov.	31	2		13	
Dec.	27	1		15	
Year	235	135		198	

EXERCISE 29

One numerical measure of variability is the relative range in percent of the mean, that is, the difference between the highest and lowest values in a given period of time divided by the average value for that period and multiplied by 100.

a) Compute the relative range for each month and for the year.

b) Why is the annual relative range so much smaller than the average of the monthly relative ranges?

c) Is the average relative range for the warmest months, May through October, smaller or larger than that for November through April? Explain.

Paradoxically, the weak storms of summer often produce the most rain over the continents at high latitudes where warm moist air is lifted along the Polar Front, while the great storms of winter rage at lower latitudes and over the seas; when winter storms do occur over the continents at high latitudes, they are quite dry.

6.2 STORMS

Storm frequency and intensity are important elements in most climates. "Storm" is an unfortunate term inasmuch as it carries a negative connotation, implying destruction or, at least, discomfort. In fact, most atmospheric circulations that are truly storms provide far more benefits than otherwise, the most obvious being rain. At the same time, some of the most damaging atmospheric events are in no sense of the word associated with storms; indeed, the lack of storms that may result in severe droughts is a good example. To complicate matters further, situations such as the heavy snows in the lee of the Great Lakes or the strong winds that occasionally sweep out of the Alps into southern France are called storms locally, but they bear little relation to the migratory atmospheric circulations which meteorologists call cyclones or storms.

Therefore, without prejudice as to nomenclature, those atmospheric features that produce clouds, precipitation, or strong surface winds either singly or in combination will be examined as climatological elements.

On a scale encompassing distances of hundreds of miles, organized upward motion of the air generates large cloud sheets and associated precipitation patterns in the vicinity of low-pressure systems. Low-pressure areas are, therefore, regions of "bad" weather; their horizontal winds exhibit a counterclockwise rotation about them on the northern

hemisphere, clockwise rotation on the southern hemisphere. These are the truly cyclonic circulations (see Figure 3.3), quite numerous on any given day and affecting some regions as often as 100 times a year.

It is known from theoretical considerations and experiment that cyclones outside the tropics, that is, extratropical cyclones, together with their complementary anticyclones (high-pressure systems), must form and develop rotational circulations in response to poleward distortions in the temperature field. The details of the development of these circulation features are only partially known; for climatological purposes it suffices to realize that there is a continuous cycle of formation, growth, and decay of cyclones and anticyclones. From the climatological point of view, it is also germane to identify regions in which cyclones are most likely to form and move.

Extratropical cyclones derive their energy from the horizontal temperature differences commonly found in middle and high latitudes. Horizontal temperature differences are largest at fronts; as a matter of fact, a front is defined as a zone of large horizontal temperature contrast over relatively short distances. Therefore, frontal zones are the most favorable regions for extratropical cyclones to form, and extratropical cyclones are, in turn, characterized by fronts within their circulations.

EXERCISE 30

The latitudinal temperature gradient is the south–north temperature difference divided by the south–north distance between the observation points. The mean latitudinal temperature gradient in units of degree Centigrade per degree latitude (one degree latitude = 60 naut. miles = 111 km) is obtained by dividing the temperature difference between equator and north pole by 90 degrees of latitude.

a) Using the data of Table 5.1, calculate this temperature gradient for the northern hemisphere in winter.

b) What is the gradient in summer?

c) Compute the latitudinal temperature gradients for each 15-degree latitude zone on the northern hemisphere in both winter and summer. At which latitude zone is the temperature gradient largest in winter? In summer?

d) At what time of the year would you expect the highest cyclone frequency to occur in Texas?

One of the favored regions for the path of extratropical cyclones is along the Polar Front that oscillates between about 30 and 70 degrees latitudes with a mean position near 50 degrees. Hence, these latitudes are a zone of relatively frequent precipitation and marked temperature changes. But it should be remembered that averages tend to mask the great variability shown by individual cases; on any given day, there may be several fronts and associated cyclones in middle latitudes, the fronts having various orientations. This is largely due to the differences in thermal properties between continents and oceans and due to the interference by mountain ranges that displace frontal positions. Hence, on the average, coastlines are also likely to become frontal zones and to have relatively high cyclone frequency. The tracks of the most vigorous oceanic storms are close to the poleward boundaries of the Gulf Stream in the Atlantic and of the Kuroshio Current in the Pacific Ocean; the Northeasters of the U. S. East Coast are a notable illustration. In a similar vein, mountain barriers can block air flow and thus produce fronts; in addition, the regions just downwind of mountains are favored breeding grounds for storms. All of these causes may affect the path of an individual cyclone in very complex ways, but the average storm tracks, such as those for the month of March in Figure 6.4, do indicate the role they play.

EXERCISE 31

a) Name three coastal areas which are major storm tracks in March.

b) List two mountainous regions near which storms tend to form.

c) Is the mean polar front at the same latitude all around the earth?

Cyclones in the tropics are usually weaker than extratropical cyclones. They do not have horizontal temperature gradients as an energy source, because temperatures in the tropics are rather uniform. There are, however, many small disturbances associated with convergent winds, mainly in the Intertropical Convergence Zone. These disturbances tend to move westward in the general easterly flow, and as they pass over given areas, generate much of the total precipitation. At many tropical stations, as much as 90 percent of the rainfall is observed during these relatively infrequent periods of intense rain, and more than half of the annual total may fall from just a few storms.

Over the warm ocean, additional energy supplied to the atmosphere by latent heat of condensation, as well as by sensible heat, can generate much more intense circulations covering a hundred miles or so. In truth, when the sea surface reaches its highest temperature, these tropical cyclones can become the most vigorous storms found in the atmosphere with surface wind speeds as high as 200 mph. When they reach land,

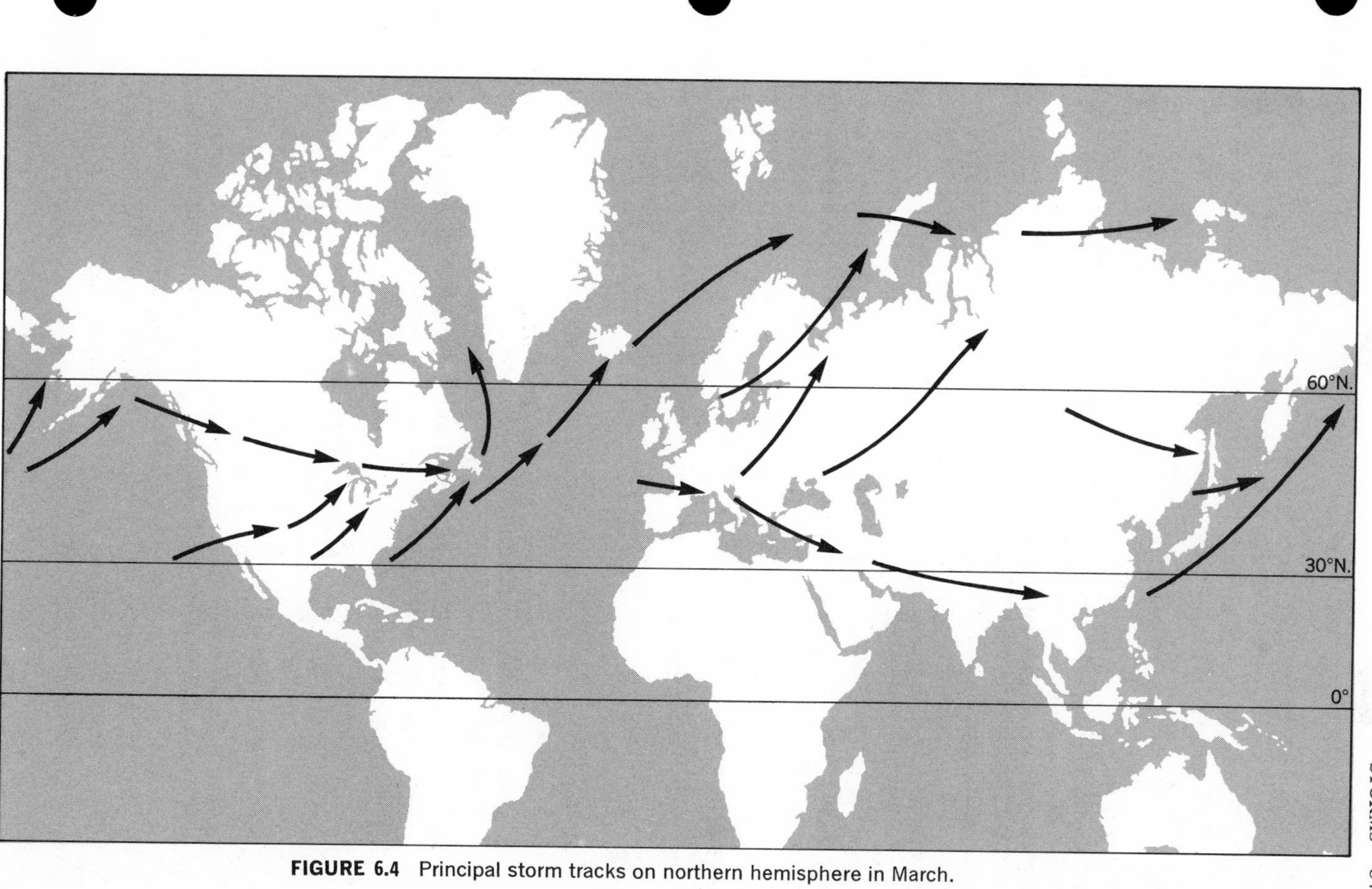

FIGURE 6.4 Principal storm tracks on northern hemisphere in March.

they lose their intensity rapidly because of greater friction and loss of the energy source. Such intense tropical storms are known by various names, the most common being hurricane in the Atlantic and typhoon in the Pacific. There is a striking correspondence between the sea surface temperature and the frequency of tropical storms as shown in Table 6.3.

TABLE 6.3
Frequency of Tropical Storms Including Hurricanes, Mean Sea Surface Temperatures for the North Atlantic at 21°N, 65°W, and Mean Rainfall at Key West, Florida

	JAN.	FEB.	MAR.	APR.	MAY	JUNE	JULY	AUG.	SEP.	OCT.	NOV.	DEC
Total Trop. Storms in N. Atlantic 1886–1964	0	1	1	0	10	41	46	142	216	146	29	4
Water Surface Temperatures (°F)	77	77	77	77	79	81	82	82	83	82	81	79
Key West, Fla. Precipitation (mm)	38	51	44	64	70	102	106	108	166	149	71	43

EXERCISE 32

a) Does the term "hurricane season" have any justification? When is that season?

b) Does the peak amount of rainfall at Key West indicate a high frequency of rainy days during that month? Explain.

c) Would you expect the maximum monthly rainfall at Key West to occur in September every year? Give your reasons.

In both the Atlantic and Pacific, tropical storms tend to invade higher latitudes under the influence of southerly flow at the western end of the subtropical high-pressure cells. Thus, the Gulf Coast, the U. S. East Coast, Japan, Taiwan, and the coast of China are susceptible to damaging winds and flooding from these storms in late summer and autumn. Indeed, the very high tides, or storm surges, that are typically found just ahead of tropical storms are their most dangerous aspect. Figure 6.5 gives an idealized picture of tracks of tropical storms throughout the world. While individual storms may follow very different paths, these average tracks reflect the influence of the mean wind patterns which have been shown in Figure 5.3.

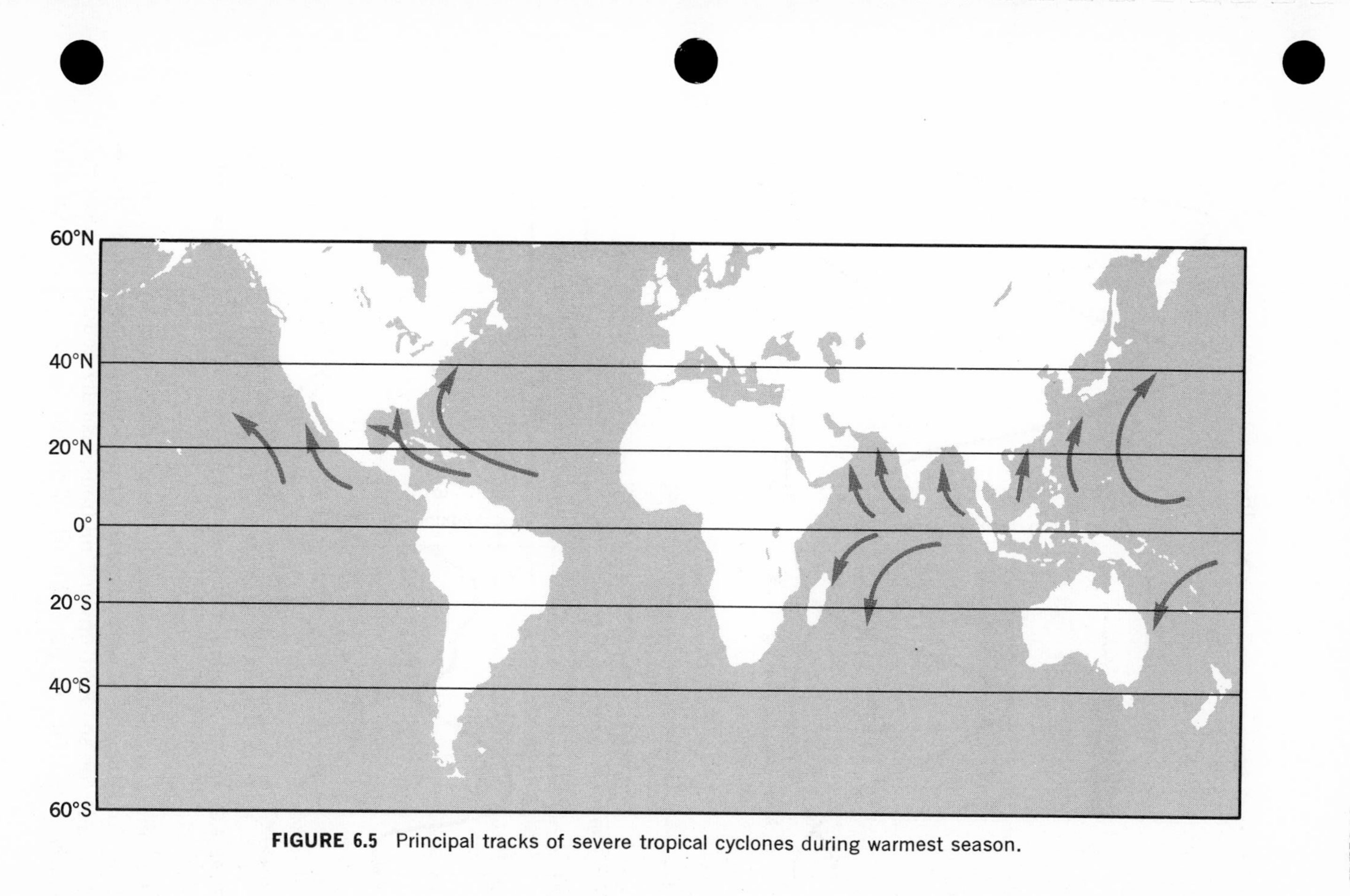

FIGURE 6.5 Principal tracks of severe tropical cyclones during warmest season.

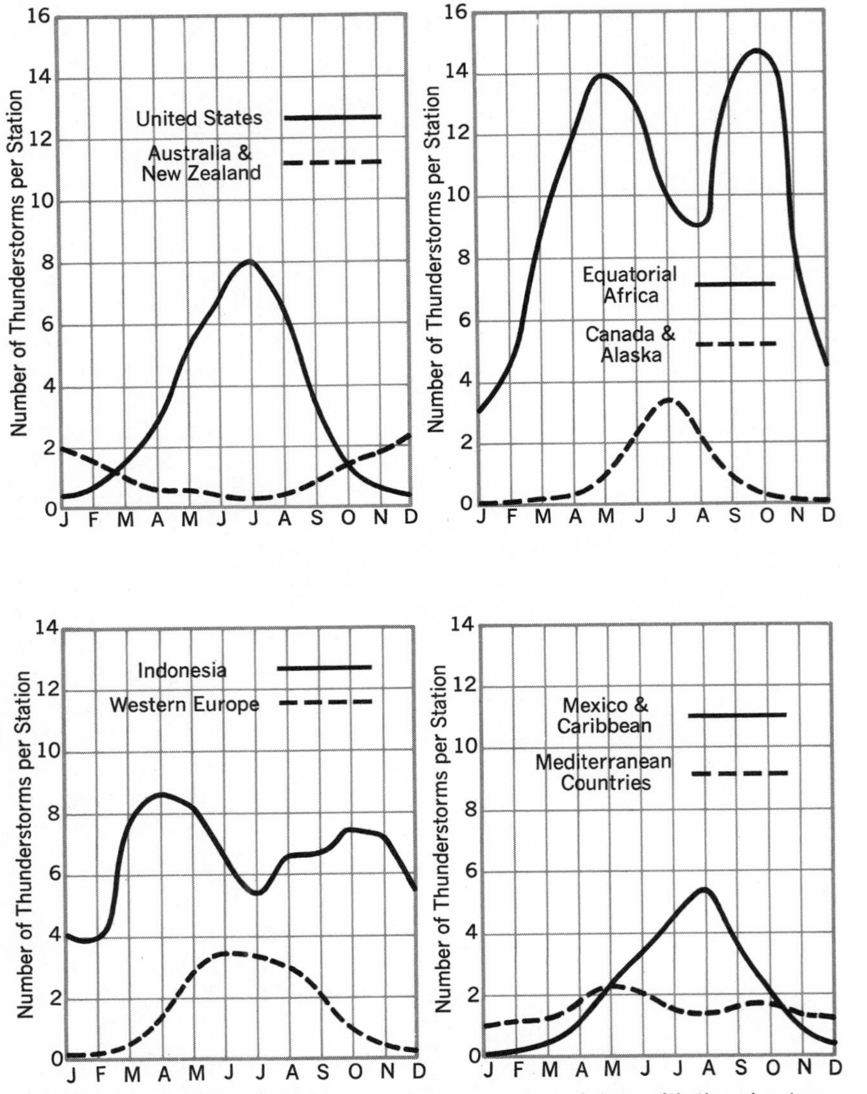

FIGURE 6.6 Annual variation of the average number of days with thunderstorms per station in selected regions.

The most common small-scale storm is the thunderstorm, which extends to a height of several miles while having a horizontal size of only a mile or so. Fast upward motions, usually caused by local differences in heating, are the hallmark of thunderstorms, and compensating downdrafts help to produce very rapid vertical exchanges in the vicinity of these cells. It is fair to visualize the thunderstorm structure as a very deep eddy, capable of exchanging slow moving air at the surface with air moving fast horizontally much higher up. When this occurs, strong, gusty winds at the ground can cause considerable damage as can the associated heavy rainshowers, hail, and lightning.

Thunderstorms, which require low-level heating, are most common in the afternoon or early evening over land and are very common in the Tropics. Further, they exhibit a pronounced seasonal shift, being very rare in the winter over high latitudes. All told, the number of thunderstorms occurring in the atmosphere on any given day probably approaches 50,000; recent satellite data have shown them to be quite numerous over the oceans, so the number quoted may be quite conservative.

Seasonal and geographical variations in thunderstorm frequency are excellent indicators of the effects of insolation as modified by the shifts in the large-scale circulation patterns. The data of Figure 6.6 provide a good review of the matters discussed in earlier chapters, as these few generalizations should show:

At high latitudes, such as in Canada and Alaska, few thunderstorms occur and then almost exclusively in summer, when the insolation is relatively strong. Equatorial Africa illustrates the double maximum typical of stations near the equator where the ICZ passes northward in spring and southward in autumn, following the sun. Mediterranean stations do not show much thunderstorm activity, as they are under the subsiding northern branch of the Hadley cell during their hot season. Finally, Mexico and the Caribbean show the late summer intrusion of the better-developed northern ICZ, with its maximum of tropical circulations.

EXERCISE 33

a) How do you account for the double maximum in thunderstorm activity in Indonesia?

b) Why does Australia have so many thunderstorms in December?

c) Which curve would best duplicate the thunderstorm frequency in California?

d) Would you expect a large total number of thunderstorms there? Explain.

e) Does the United States have a "monsoonal" thunderstorm distribution related to the heating of the continent or one related to heat storage in the ocean? On what basis did you choose?

The relative importance of thunderstorms as an element of individual climates is a useful key to the temperature structure of the atmosphere and thus to the observed precipitation–cloudiness relationship. That is to say, where precipitation is predominantly produced by cyclonic activity, cloudiness and precipitation are usually at a maximum at the same time. On the other hand, locations that receive most of their rainfall from showers and thundershowers often have an inverse relationship between cloudiness and precipitation due to the sinking cloud-free air near the thunderstorm cells. It is difficult to generalize about the timing of maximum cloudiness in relation to maximum precipitation, but usually when both maxima occur in the same month, large-scale cyclones are more significant than thunderstorms in causing the precipitation.

The data from Rome, Italy, and Strasbourg, France, are a case in point. It is interesting to note in Table 6.4 that Rome has less cloudiness but more precipitation than Strasbourg; further, in Table 6.5 the precipitation and cloudiness are well related at Rome, where thunderstorm activity does not vary greatly with season, but not well related at Strasbourg, where the seasonal variation of thunderstorms is large with a prominent peak in summer.

TABLE 6.4

Annual Mean Cloudiness, Total Precipitation, and Mean Number of Thunderstorm Days at Rome, Italy, and Strasbourg, France

	CLOUDINESS (percent)	PRECIPITATION (mm)	THUNDERSTORMS (no.)
Rome	47	711	17
Strasbourg	65	607	18

TABLE 6.5

Percentage of the Annual Average Cloudiness, Precipitation, and Thunderstorm Frequency at Rome and Strasbourg by Seasons

	ROME			STRASBOURG		
	CLOUDINESS (percent)	PRECIPITATION (percent)	THUNDERSTORMS (percent)	CLOUDINESS (percent)	PRECIPITATION (percent)	THUNDERSTORMS (percent)
Spring	28	24	29	23	21	25
Summer	16	10	27	23	39	54
Autumn	25	34	25	25	23	19
Winter	31	32	19	29	17	2
Year	100	100	100	100	100	100

EXERCISE 34

a) Discuss the seasonal contrasts in cloudiness, precipitation, and thunderstorm frequency at the two stations.

b) Regarding the seasonal variation of precipitation, which of the stations in Table 6.1 does Rome resemble most closely, and in what way is there a difference between these stations?

Thunderstorms always have a certain randomness about them; local variations are large. However, thunderstorms are always more numerous within the circulation of cyclones, both tropical and extratropical. This may seem contradictory, inasmuch as cyclones are large-scale phenomena and often cause widespread rains, but it is also true that many cyclones are little more than regions of increased shower activity, generally when the low-level air is warm.

Occasionally, tornadoes, those very small but extremely destructive vortices, form in a few of the most intense thunderstorms. Tornadoes are very much more common in the U. S. than elsewhere in the world, a rather unhappy accident of a combination of geographic and meteorological features. What is usually required for their formation is a simultaneous arrangement of warm moist air near the surface, cold dry air aloft, and a rather vigorous polar front in the vicinity. In contrast to other middle-latitude locations, the U. S. has the excellent source of low-level warm moist air from the Gulf of Mexico, desiccated air aloft flowing off the Rockies, combined with very cold air surging out of Canada. Finally, hurricanes occasionally produce tornadoes, but these may have a somewhat different mode of origin and are a small fraction of the U. S. total.

To the extent that the factors enumerated above promote tornadoes, it should be expected that they would be most frequent in the spring when the cold air from Canada is still markedly cold, while the Gulf air has already taken on summer-like characteristics. This can be verified from Table 6.6.

TABLE 6.6
Monthly Percentage Distribution of Thunderstorm Days and Tornadoes in the United States 1916–1964

Month	Jan.	Feb.	Mar.	Apr.	May	June	July	Aug.	Sep.	Oct.	Nov.	Dec.
Percentage of Annual Thunderstorm Days	1.1	1.6	4.0	7.0	13.3	17.6	21.3	17.8	9.9	3.7	1.6	1.1
Percentage of Annual Tornadoes	2.1	3.4	9.8	15.7	22.9	18.2	9.8	5.5	4.7	2.5	3.4	2.0

EXERCISE 35

Plot the data on the graph of Figure 6.7.

a) Why does the tornado maximum not occur in the same month as the thunderstorm maximum?

b) When is the tornado season?

c) Do the data show a strong hurricane influence on tornado frequency? Explain your answer.

The distribution of tornadoes within the U. S. is quite interesting and suggests that they are most common over smooth terrain. The Rocky Mountain States, the Appalachian region, and the Pacific Coast states all have relatively low tornado frequencies, while the states in the Plains and along the East and Gulf Coasts have higher frequencies. Table 6.7 gives the relevant data, including the frequency per 1000 square mile area for a recent 12-year period. Some striking oddities occur when the reduction to a standard area is made, notably the very high frequency in Massachusetts and, to some extent, Delaware. In the case of Delaware, the small size of the state is probably responsible, and Massachusetts' value may be distorted, but the data do suggest that a secondary maximum is present in New England. The statistics of infrequent climatic events are notoriously misleading, especially as they depend heavily on fortuitous reporting. However, these data are from a recent period, when reporting was improved, and are verified occurrences. In any event, to assert that tornadoes pose the greatest threat in Texas merely because more have been reported from that large state is unjustified.

EXERCISE 36

a) Rank the ten top "tornado" states in terms of their frequency per 1000 square miles.

b) Which is more likely to have a tornado, a town in Alaska or the same size town in Rhode Island?

c) Compare the climatological significance of tornadoes with that of hurricanes.

Finally, a word should be said about the many locations on the earth having strong and sometimes damaging winds that are local in nature. In nearly every case, these are related to downslope flows near mountainous terrain, and many of them have picturesque local names. The

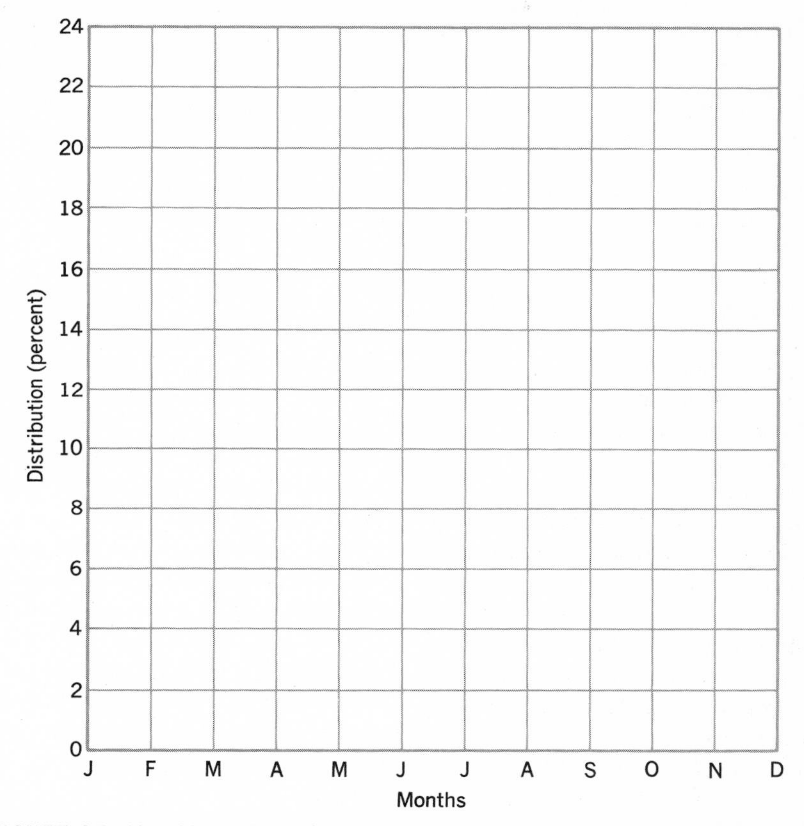

FIGURE 6.7 Monthly percent distribution of tornadoes and thunderstorm days in the United States 1916–1964.

TABLE 6.7
Total Number of Tornadoes per State and Total Number of Tornadoes per 1000 Square Miles per State for the Period 1953–1964

State	Tornado Total	Tornadoes per 1000 Square Miles	State	Tornado Total	Tornadoes per 1000 Square Miles
Texas	1074	4.1	Pennsylvania	72	1.6
Oklahoma	799	11.5	Kentucky	59	1.5
Kansas	721	8.8	Massachusetts	57	7.2
Nebraska	435	5.7	Virginia	52	1.3
Missouri	308	4.5	Montana	43	0.3
Illinois	274	4.9	Maine	40	1.3
Florida	262	4.8	Arizona	28	0.2
Indiana	252	7.0	New Hampshire	27	3.0
Iowa	236	4.2	California	26	0.2
Alabama	215	4.2	New York	24	0.5
Arkansas	212	4.0	Maryland	22	2.2
Georgia	205	3.5	Idaho	19	0.2
South Dakota	181	2.4	New Jersey	19	2.5
Louisiana	177	3.9	Vermont	17	1.8
Wisconsin	162	3.0	Connecticut	16	3.3
Mississippi	162	3.4	Utah	10	0.1
Colorado	158	1.5	Washington	9	0.1
Minnesota	144	1.8	Delaware	9	4.6
Michigan	121	2.1	West Virginia	9	0.4
North Dakota	121	1.7	Oregon	5	0.1
South Carolina	109	3.6	Nevada	5	0.02
Ohio	108	2.6	Hawaii	3	0.47
New Mexico	99	0.8	Alaska	1	0.002
North Carolina	95	1.9	Rhode Island	0	0
Tennessee	80	1.9	District of Columbia	0	0
Wyoming	72	0.7			

westerly Chinook in the Rocky Mountains is a warm dry wind which often reaches high speeds. An analogous east wind from the mountains in California is called the Santa Ana; it is hot and dry and often associated with serious brush fires in Southern California.

When very cold air is dammed up by mountains, it may eventually spill down into the warm low-lying areas nearby. Although the air warms up by compression during descent, when it arrives in the low elevation at very high speed it is still colder than the air that had been there. Cold winds of this sort are called the Bora along the Adriatic Coast and the Mistral in Southern France. There is a similar wind along the Alaskan coast, which rushes down onto the coastal plain at speeds up to one hundred miles per hour. Countless local situations induced by channeling effects on the wind or lake effects are called storms by those exposed to them, and it is difficult to quarrel with the usage.

6.3 IRREGULAR TEMPERATURE AND HUMIDITY CHANGES

In middle latitudes, migratory lows and highs transport cold air from polar regions and warm air from tropical regions; the leading surfaces of these advected air masses are called cold fronts and warm fronts, respectively. When these fronts pass a station, they produce aperiodic and sometimes sudden changes in temperature, dew point, and relative humidity that can disrupt the normal diurnal variations. In addition, because of the convergence and lifting of air along the fronts, they are often accompanied by cloudiness and precipitation. In Table 6.8 examples of both kinds of fronts are given as observed at University Park, Pennsylvania, together with days exhibiting the usual diurnal variations.

TABLE 6.8
Effects of Warm and Cold Front Passages on the Diurnal Variations of Temperature, Dew Point, and Relative Humidity

TIME	2/3 MAY 1967			31 JAN./1 FEB. 1967			2/3 JUNE 1967		
(hours)	TEMP. (°F)	DEW PT. (°F)	REL. H. (percent)	TEMP. (°F)	DEW PT. (°F)	REL. H. (percent)	TEMP. (°F)	DEW PT. (°F)	REL. H. (percent)
00	67	61	80	23	15	71	56	41	57
02	67	64	90	21	18	88	53	42	66
04	63	61	93	17	16	92	49	43	79
06	60	58	94	15	13	92	49	44	83
08	63	61	93	14	13	93	61	44	53
10	65	59	82	22	16	77	74	45	36
12	72	60	65	29	16	53	78	43	29
14	70	70	100	32	18	52	81	43	27
16	64	64	100	31	30	94	82	44	26
18	53	52	96	31	31	100	81	44	27
20	52	48	87	33	33	100	73	44	36
22	51	47	88	33	33	100	67	45	45
00	50	48	94	35	35	100	65	49	58
02	47	41	81	39	39	100	61	51	70
04	45	40	80	40	38	93	57	51	80
06	44	39	80	39	37	93	57	51	80
08	45	29	53	38	36	93	65	50	60
10	49	23	35	37	36	96	76	48	37
12	52	25	34	38	36	93	79	44	29
14	56	24	28	39	34	85	82	46	28
16	58	23	25	39	34	85	83	45	27
18	57	22	25	38	33	84	79	43	28
20	51	22	30	37	33	85	73	42	33

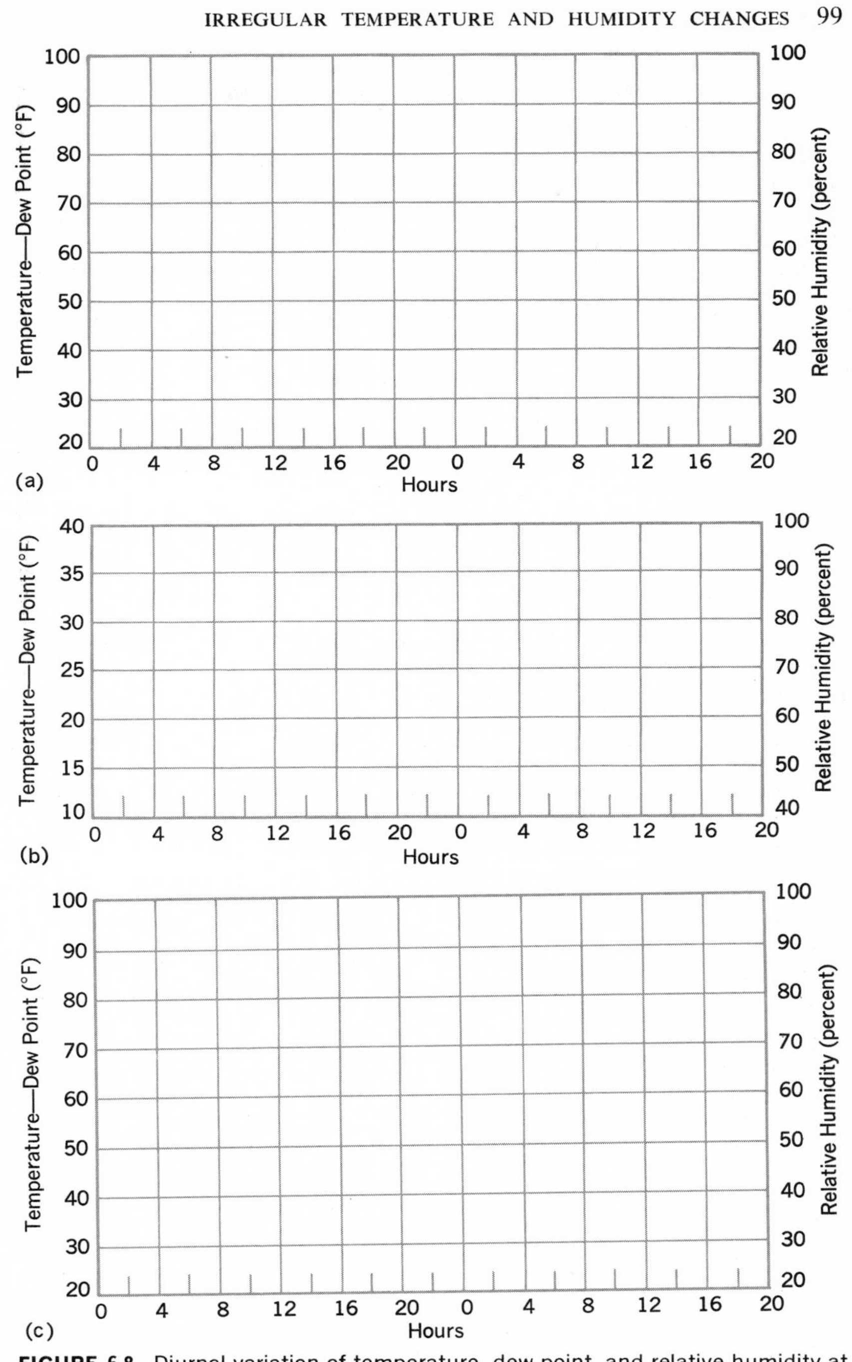

FIGURE 6.8 Diurnal variation of temperature, dew point, and relative humidity at University Park, Pa., (a) on 2/3 May 1967; (b) on 31 January and 1 February 1967; (c) on 2/3 June 1967.

EXERCISE 37

a) Determine the mean temperatures for each of the four days as the averages of the respective highest and lowest temperatures of each. The change of the mean temperature from one day to the next is called interdiurnal temperature change. By means of the sign of the interdiurnal temperature changes, distinguish the warm-front case from the cold-front case.

b) Plot the data from Table 6.8 on the graphs of Figure 6.8 (a), (b), and (c), respectively. What causes the sudden rise in dew point, and, especially, relative humidity in the afternoon of 2 May and 31 January?

c) Which quantity, relative humidity or dew point, more clearly shows the changes in vapor content of the air after the passage of these fronts?

In summary, in regions of frequent front passages, one can expect large and rapid variations of climatic elements; in particular, the interdiurnal variation of temperature will be large, especially in winter and spring, when the cyclones and anticyclones are best developed and have the highest frequency of occurrence. It should also be noted that the oceans moderate the temperature contrasts of opposing air masses and reduce the interdiurnal temperature changes observed in maritime climates as compared to those in continental climates at the same latitudes. Thus, a large average interdiurnal temperature change is a mark of strong continentality, as well as of frequent passages of fronts.

6.4 CLOUDS AND FOGS AS CONTROLS

Much of the energy transfer from the earth's surface to the air is by long-wave radiation (infrared) and by the latent heat of vaporization. The condensation process produces clouds which, in turn, reduce incoming energy from the sun and sky chiefly by reflection back into space. In addition, clouds absorb and reradiate the outgoing terrestrial radiation back to the earth. In general, they moderate the maximum temperature in daytime and the minimum temperature at night. Cloudiness effects are illustrated by the average diurnal temperature variations shown in Figure 6.9; although the average temperatures for the two sets of days are the same (55°F), the diurnal range of temperature on the clear days is four times as large as that on the overcast days.

EXERCISE 38

a) What is the effect of cloudiness on the "continentality" of a station?

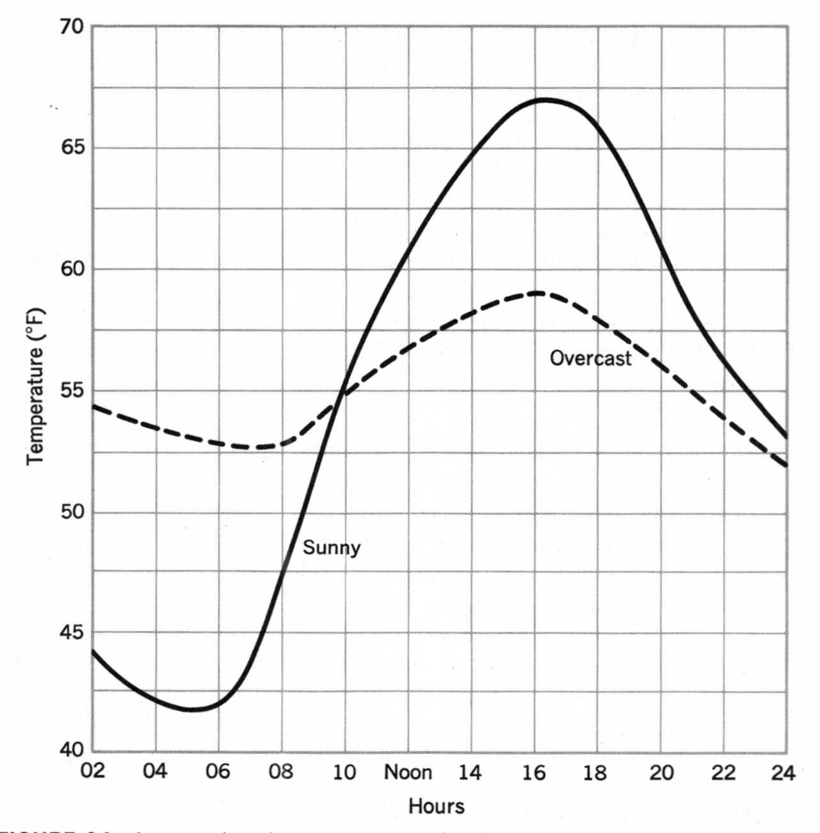

FIGURE 6.9 Average hourly temperatures for six clear and six overcast days between March and June 1968 at University Park, Pa. (The daily averages for each set are the same at 55°F.)

b) If you have no direct information on cloudiness, how could you estimate qualitatively the difference in cloudiness between March and September at a given station?

In addition to the direct effects on temperature associated with an outbreak of polar air or an invasion of tropical air, the changed cloudiness after the passage of a front may exaggerate or mitigate the temperature changes that would otherwise be expected. For example, in Figure 6.8(b) the air would be a good deal warmer on 1 February if it were not for persistent cloudiness. Similarly, if the daytime hours of 3 May in Figure 6.8(a) had been cloudy, that day would have been colder.

On a longer time scale, cloudiness may also affect the annual march of

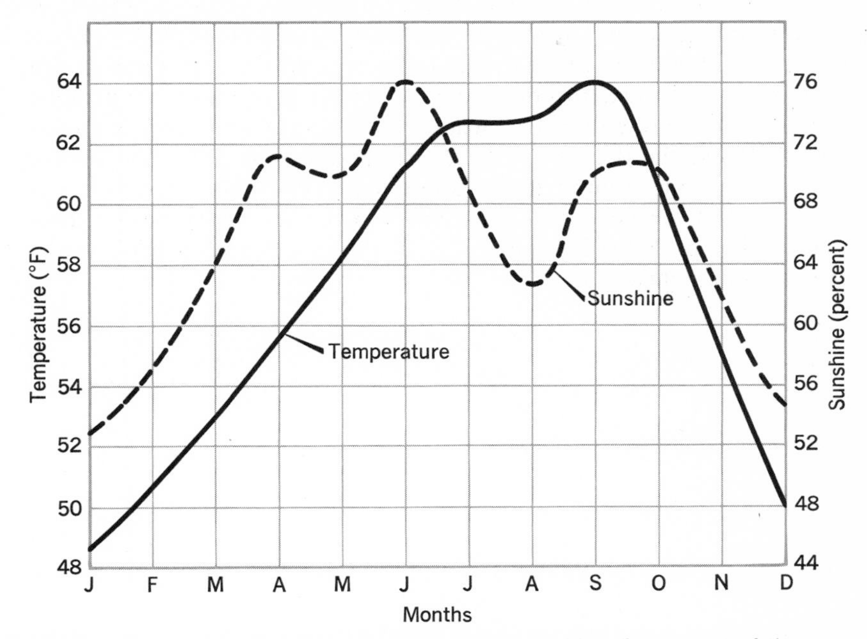

FIGURE 6.10 Mean monthly temperatures and sunshine (in percent of the possible sunshine) at San Francisco, California.

FIGURE 6.11 Aerial view of fog layer beyond Golden Gate Bridge in San Francisco. (Courtesy Dr. Albert Miller, Professor of Meteorology, San Jose State College.)

temperature, if extensive cloudiness prevails during a particular season or if there is a season in which fog frequency is especially high. This is demonstrated in Figure 6.10 by the mean monthly temperatures at San Francisco, California, together with the mean percentages of possible sunshine which are an inverse measure of the combined effect of cloudiness and fog; the depression of the temperature in July and August, which delays the annual temperature maximum until September, is clearly due to the increased cloudiness and fog frequency during these months. The photograph in Figure 6.11 shows a solid fog bank in the San Francisco area.

7 effects of elevation and surface characteristics

7.1 VERTICAL TEMPERATURE GRADIENTS

On the average, the temperature of the lower free atmosphere decreases upward for several miles, because the air is primarily heated by the earth's surface; in addition, temperature also decreases upward when measured at ground levels of different elevations. This can be seen from Table 7.1 in which the mean temperatures are presented for various altitudes at O'Neill, Nebraska, for August and September 1953 and for the U. S. Standard Atmosphere; the latter is derived as an average from many U. S. stations over long periods. The fact that the temperature decreases with height along the ground also is shown by the average data for Switzerland, California, and Ethiopia.

TABLE 7.1
Mean Temperatures at Various Altitudes for Different Places

Free-Air Observations				Surface Observations					
O'Neill, Neb.		U. S. Std. Atm.		Switzerland		California		Ethiopia	
Height	Temp.	Height	Temp.	Height	Temp.	Height	Temp.	Height	Temp.
(m)	(°C)	(m)	(°C)	(m)	(°C)	(m)	(°C)	(m)	(°C)
2430	15.1	3000	−4.5	2496	−1.9	2140	6.7	2408	17.2
2130	16.8	2000	2.0	569	8.6	1561	10.2	1903	20.6
1820	18.7	1000	8.5			1180	12.0	916	26.8
1365	21.2	0	15.0			610	15.2		
1060	22.6								
910	23.3								
Average Lapse Rates in °C/100 m									

EXERCISE 39

Plot the data on the graph of Figure 7.1 and draw the best fitting straight line through each of the five sets of points.

a) Determine the slope of these lines in terms of °C/100 m; these are the average "lapse rates," found by taking two convenient points, one each near the ends of the line, and dividing the differences in temperature at these points by the respective altitude differences expressed in units of 100 m.

b) Is there a significant difference between the lapse rates in the free atmosphere and those observed at the ground at different elevations?

Although the temperature levels in the various situations are quite different, the average vertical temperature gradients are remarkably similar. They are also very much greater than the average horizontal temperature gradients.

EXERCISE 40

a) From Table 5.1 compute the average horizontal (latitudinal) temperature gradient for the earth as a whole by subtracting the average of the four pole temperatures from the equator temperature and dividing by the distance between equator and poles which is 10,000 km; convert this horizontal temperature gradient into the same units as the lapse rates and compare with the lapse rates determined in Exercise 39.

b) On the average, what distance would you have to travel poleward in order to experience the same temperature drop as you would by going 200 m upward?

The fact that the lapse rate of temperature as measured along sloping terrain averages to very nearly the same as that in the free atmosphere is, to a large extent, due to ventilation of the slopes by the wind and mixing of the air near the slopes with air from the free atmosphere. The same argument does not necessarily apply to a high plateau or a mountain when the winds are light. Here the surface temperatures and the temperatures of the air near the surface are controlled by the radiation balance. Interestingly enough, the incoming solar radiation is greater as a result of less depletion, but the outgoing terrestrial radiation is also greater because there is usually less blanketing water vapor in the air at higher elevations. The large losses of infrared energy to space usually result in lower temperatures.

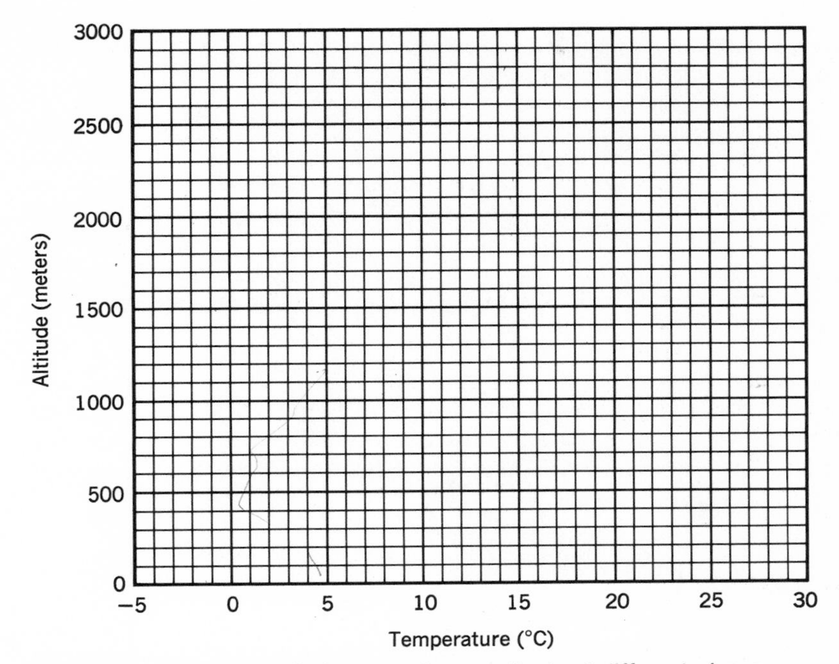

FIGURE 7.1 Average vertical temperature gradients at different places.

While temperature normally decreases with height, wide variations of lapse rates are commonly observed. The temperature at ground level is strongly controlled by the temperature of the surface itself, while the layers aloft are much less affected. As a result, the lapse rate in the lower layers of the atmosphere undergoes a regular diurnal variation with the largest lapse rates in the afternoon and small lapses of temperature or even increases of temperature with height, called inversions, during the night. Air is said to be stable with small lapses and inversions.

The stabilizing effect of air cooling near the ground is simulated in heated rooms during winter, in which an inversion often develops through the accumulation of warm air near the ceiling, where the temperature may be several degrees higher than at floor level. When someone smokes in such a room, the stability of the air can sometimes be seen by the horizontal spreading of the smoke. Outdoors, the establishment of inversions is noticeable in the evening when the smoke from backyard grills or trashburners rises several feet but then spreads out under an invisible ceiling.

More important from a climatic viewpoint are the nocturnal inversions during the colder seasons, when the temperatures at low-lying places may be well below freezing, whereas slightly higher locations may have considerably higher temperatures. As a matter of fact, some inversions exhibit such sharp temperature rises with height that the blossoms on a fruit tree may be killed by frost in the lower half of the tree, while they survive in the upper half.

Conversely, in the afternoon, strong heating at the surface leads to high temperatures and thus to large temperature lapses with height. This case is called unstable, because the heated air is very easily exchanged with cooler air from aloft. The evidence for vigorous vertical mixing is abundant: Increased visibility as atmospheric suspensions are lifted away from the ground, towering daytime clouds, and afternoon thundershowers are but three examples.

EXERCISE 41

a) In a region that has substantial air-pollution sources, at what time of the day would the concentration of pollutants near the surface be a minimum?

b) At what time would it be a maximum? Explain.

7.2 VERTICAL WIND GRADIENT

Normally, the wind increases with height, both in the free atmosphere and where the ground itself is elevated. This is illustrated for the free atmosphere by the average wind speed in knots (1 knot = 1.15 mph) at various heights on four January days in 1967 at Dunkirk, N. Y. See Table 7.2. It should be noted that this is a case of particularly strong increase of wind speed with height, that is, a strong wind shear.

TABLE 7.2
Average Wind Speeds Aloft in January 1967 at Dunkirk, N. Y.

HEIGHT (1000 ft)	0	1	2	3	4	5	6	7	8	9	10
WIND SPEED (knots)	8	19	23	24	25	27	30	35	40	45	48

EXERCISE 42

Plot the wind speeds on the graph of Figure 7.2.

a) Between which consecutive altitudes is the wind shear greatest?

b) What is the average increase in wind speed with height (in knots per 1000 ft) in the layer between 0 and 2000 ft? What between 2000 and 10,000 ft?

c) Suppose there were a 10,000-ft peak near Dunkirk, N. Y., would you expect the wind speed on the peak to be 48 knots? More? Less? Explain your answer.

That higher ground elevations are windier is evident from the average annual wind speed of 7.5 knots at Albany, N. Y., in a valley of 280 ft elevation, and the speed of 8.7 knots at Binghamton, N. Y., on an exposed hill top at 1600 ft. However, it can also be seen that the wind shear along the ground is quite small and, in general, is not as large as that in the free atmosphere, a fact that can be attributed to frictional drag.

Diurnally, the wind speed is subject to a variation that is related to the diurnal variation of the lapse rate. During the warm part of the day, when the lapse rate is unstable, air brought down from aloft preserves some of its higher horizontal speed; this results in higher average wind speed and more gustiness at the ground. In turn, the wind aloft is somewhat slowed by the rising air from the surface. Conversely, at night, when the atmosphere is stable, there is little coupling between the air near the surface and air layers aloft, with the result that nighttime winds are stronger aloft and very weak near the surface. This is shown by the average diurnal wind-speed variations at 37- and 410-ft height

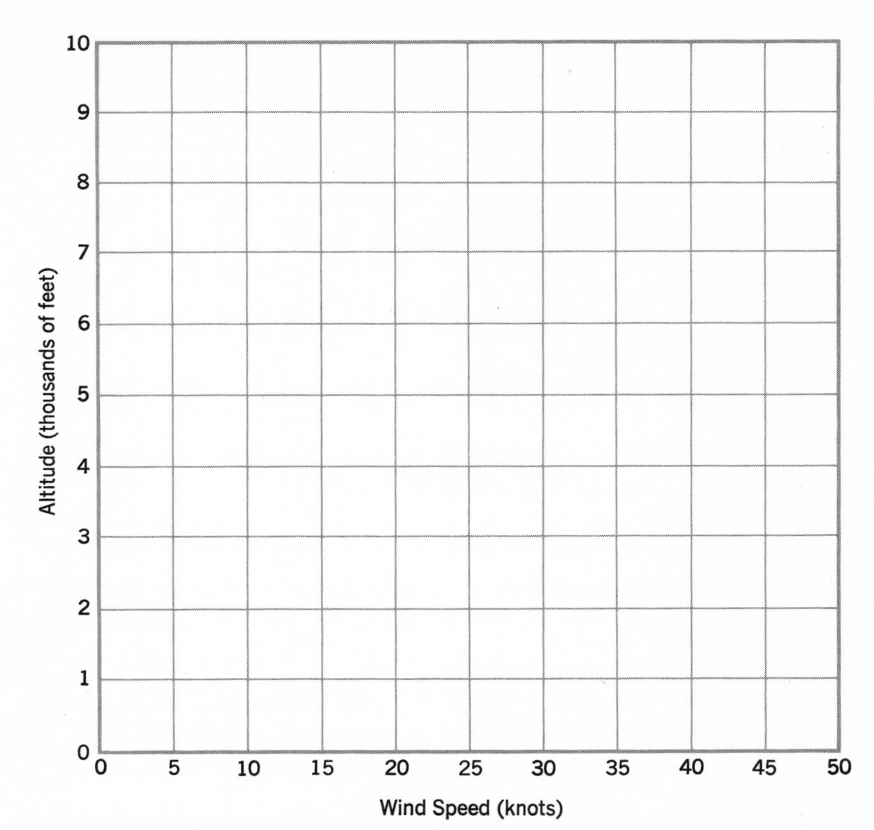

FIGURE 7.2 Vertical distribution of wind speed over Dunkirk, N. Y., in January 1967.

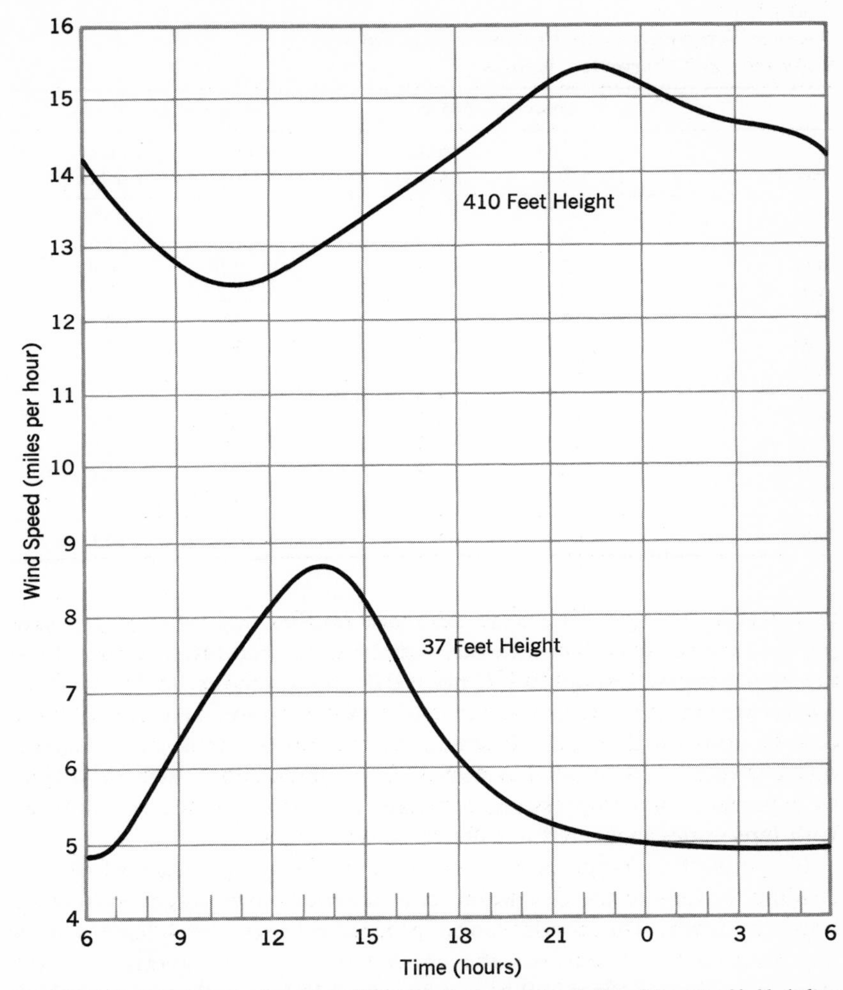

FIGURE 7.3 Two-year averages of hourly wind speeds at Brookhaven, N. Y. (after I. A. Singer and G. S. Raynor).

above Brookhaven, N. Y., in Figure 7.3. There, we see that the low-level speed increases to a maximum at about 14 hours, while the drag exerted by the lower air on the air flow at 410-ft height causes a minimum around noon and delays the maximum until 22 hours.

7.3 TOPOGRAPHIC EFFECTS

Mountains present obstacles to the flow of air, forcing it to ascend on the windward side and to descend on the lee. The temperature changes of air resulting from its compression or expansion alone are called adiabatic

TABLE 7.3
Average Temperature and Precipitation at Denver, Colorado, and Marysville, Kansas

	Denver, 1615 m		Marysville, 370 m	
Month	Temp. (°C)	Precip. (inches)	Temp. (°C)	Precip. (inches)
Jan.	−0.3	0.5	−4.1	0.9
Feb.	0.7	0.9	−1.1	1.3
Mar.	2.9	1.4	2.6	1.9
Apr.	8.1	1.8	11.2	2.9
May	14.1	3.0	17.4	4.4
June	20.2	1.0	23.2	4.6
July	23.1	2.0	25.8	4.6
Aug.	22.3	1.5	25.7	4.3
Sep.	17.8	0.9	20.6	3.4
Oct.	11.3	1.2	13.6	1.8
Nov.	3.7	0.8	4.4	1.2
Dec.	0.9	0.6	−0.6	0.8
Year	10.4	15.6	11.6	32.1

temperature changes. The word adiabatic implies that no energy is supplied to nor removed from the air. Adiabatic temperature changes of rising or sinking air amount to 1°C per 100 m altitude change ($5\frac{1}{2}$°F/1000 ft), a decrease in temperature in the case of upward motion, an increase in the case of downward motion. It should be emphasized that such temperature changes occur in air that moves up or down and is thereby subject to expansion or compression, respectively, and must not be confused with lapse rates measured in different layers of air.

If a mountain range is sufficiently high, the temperature of the ascending air falls to the dew point, condensation takes place, and clouds begin to form. The condensation process releases the latent heat of vaporization which opposes the cooling due to the expansion of the rising air. The net effect will be a reduced cooling rate, that is, one which is less than 1°C/100 m, depending on the amount of water condensed. With sufficient cloud development, precipitation will fall out of the clouds thereby removing some of the water content of the ascending air.

After the air has passed over the mountain crest, it begins its descent which will start the air temperature rising, essentially at the adiabatic rate. Because of the release of latent heat and removal of water by precipitation on the windward side, the air will arrive warmer and drier on the lee than it was at the same level on the other side of the mountains. This warm, dry, and often gusty wind coming down on the lee of mountains is called the Chinook or Foehn.

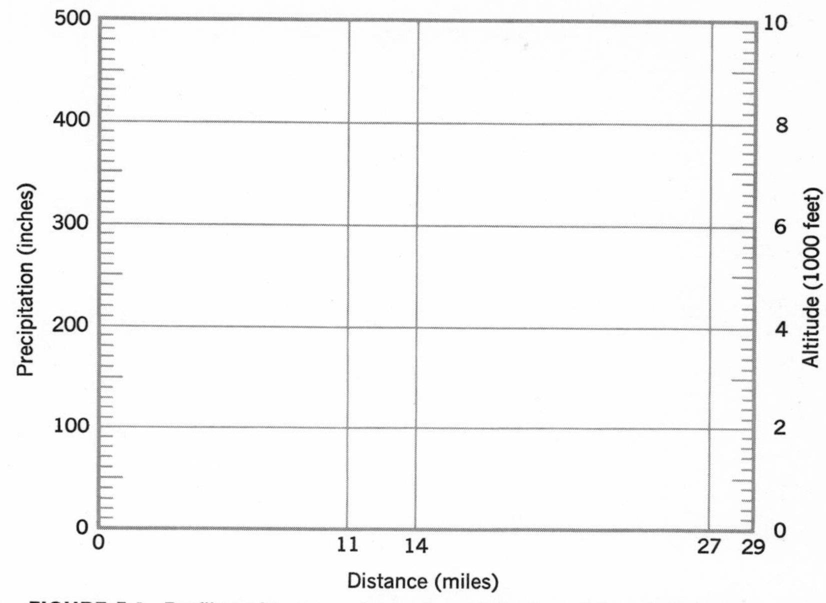

FIGURE 7.4 Profiles of topography and precipitation of Kauai, Hawaii.

The Chinook effect can represent a strong climatic control where mountain ranges are more or less perpendicular to the prevailing wind. For its altitude, Denver, Colorado, for example, appears to have too high a temperature and too little precipitation as compared to Marysville, Kansas, which is at about the same latitude, but 380 miles east of Denver, largely outside the influence of the Rocky Mountains. (See Table 7.3.)

EXERCISE 43

a) Using the mean annual temperature at Marysville and the standard lapse rate of 0.6°C/100 m, estimate the temperature at 1245 m above Marysville, which would be at the height of Denver. How much warmer does Denver turn out to be?

b) What indirect Chinook effect also contributes to the high temperature at Denver?

One of the most dramatic topographic effects on climate can be found in precipitation data; generally, precipitation increases with elevation, but as the Denver–Marysville example shows, there are many exceptions. This is especially true at the tops of very high mountain ranges; there, the air is usually quite dry. A particularly striking example of the more common situation can be drawn from Kauai Island of Hawaii where an ENE–WSW cross section from one end of the island to the opposite end shows what is probably the greatest precipitation gradient on earth. Not only is the rainfall increase with altitude revealed in Table 7.4, but also the great contrast between opposite sides of the island.

TABLE 7.4
Average Annual Precipitation Profile through Kauai, Hawaii

Station	Distance (miles)	Altitude (feet)	Precipitation (inches)
Anahola	0	40	52
Hanalei Tunnel	11	1218	175
Mt. Waialeale	14	5080	465
Hukipo	27	800	22
Kekaha	29	9	20

EXERCISE 44

Draw the profiles of altitude and precipitation on the graph of Figure 7.4.

a) Which station is located at the ENE end of the island? Explain.

b) Determine the approximate horizontal and vertical precipitation gradients between Anahola and Mt. Waialeale in units of inches per mile; do the same for Kekaha and Mt. Waialeale. Which gradients are larger, the horizontal or the vertical ones?

A single mountain range causes a general increase in precipitation with height. The details of precipitation and cloudiness distribution over a series of successive chains of mountains become rather complex, because the amount of precipitation is sensitive to the available water vapor

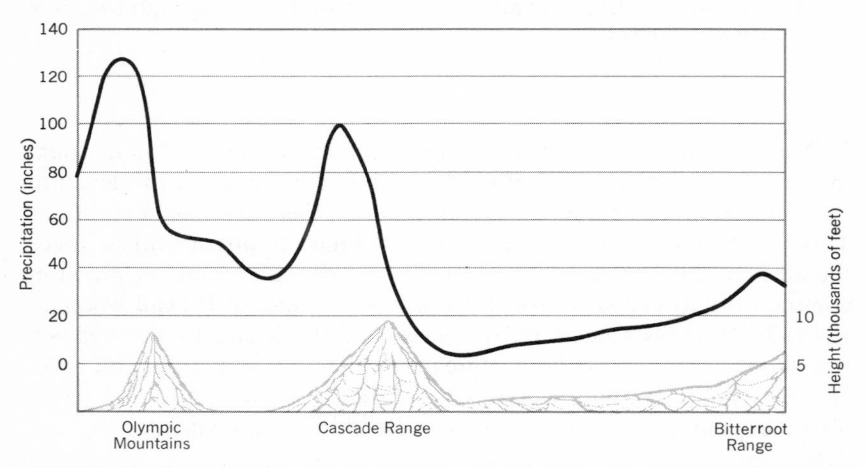

FIGURE 7.5 Topographic and precipitation profiles from the Olympic Mountains in Washington to the Bitterroot Range in Idaho.

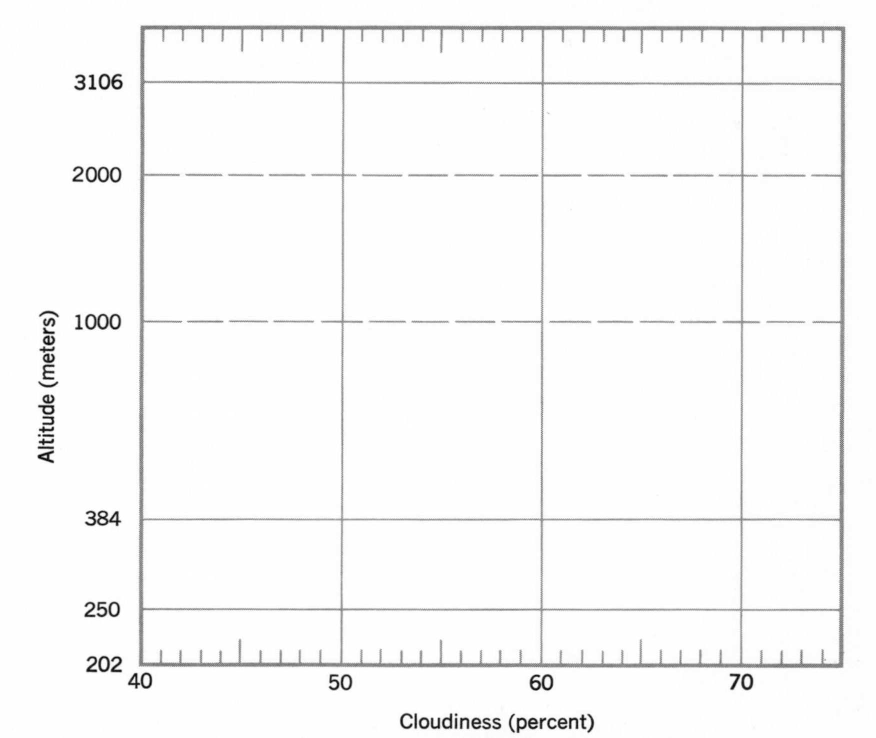

FIGURE 7.6 Change of cloudiness with altitude in Austria. Note that the ordinate scale is logarithmic in order to spread out the small differences of the lower altitudes and contract the large span to the one high altitude.

supply; hence, while upwind slopes usually receive more rain than downwind slopes, precipitation diminishes over successive slopes. This is illustrated in Figure 7.5 by the topographic and precipitation profiles in a cross section through the Olympic Mountains and the Cascade Range in Washington to the Bitterroot Range in Idaho.

As was already evident from Figure 4.2, there is no simple relationship between precipitation and cloudiness. Although precipitation usually increases with height, this is true for cloudiness in only very few regions. In Table 7.5 the mean cloudiness for four Austrian stations is given for winter, summer, and the whole year.

TABLE 7.5
Change of Cloudiness with Altitude in Austria

Station	Altitude (meters)	Cloudiness (percent) Winter	Summer	Year
Vienna	202	70	48	58
Grein	250	72	52	61
Kremsmünster	384	74	57	65
Sonnblick Mountain	3106	54	72	63

EXERCISE 45

Plot the data on the graph of Figure 7.6. At roughly what altitude does the seasonal variation of cloudiness reverse itself in this case?

For the lower altitudes cloudiness does increase slightly with height, but Sonnblick Mountain has a seasonal distribution reverse from that of the other stations. This reversal is, in part, due to lower clouds in winter that leave the mountain peaks in sunshine, whereas summer clouds resulting from greater heating frequently envelop the highest peaks but let sunshine through to the valleys and plains.

7.4 LOCAL EFFECTS OF CONTOURS

On a much smaller scale, local variations in altitude and roughness can produce notable climatic differences over distances of the order of a few miles or less. Such differences with regard to precipitation are usually greater with shower-type precipitation than with cyclonic precipitation, because the life time and size of thunderclouds are limited. Moreover, formation, dissipation, and paths of thunderstorms are strongly influenced by surface features such as hills and rivers so that systematic differences in rainfall may be recorded at various stations in the same

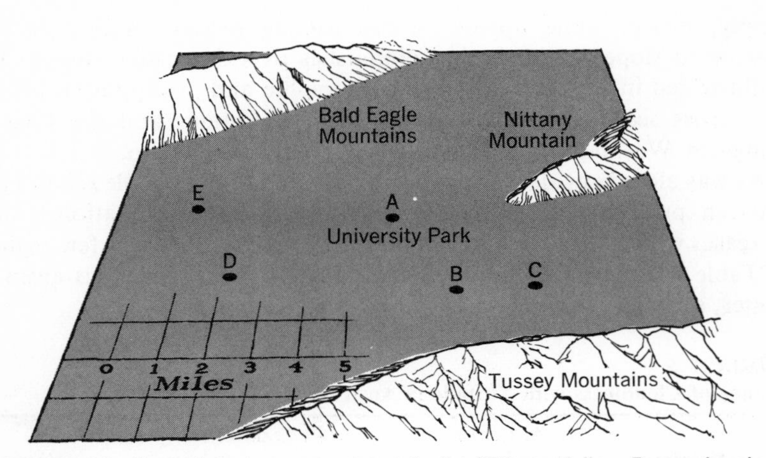

FIGURE 7.7 Part of a rain gage network in the Nittany Valley, Pennsylvania.

area. This is shown by the data collected in a special rain gage network in the Nittany Valley surrounding University Park, Pennsylvania, designated as Station A in Table 7.6; four other auxiliary stations with their respective distances from Station A are also given. The location of the various stations is shown on the map of Figure 7.7.

TABLE 7.6
Precipitation (inches) in the Nittany Valley during Summer 1965 (After E. G. Reich)

Station	Distance (miles)	June	July	Aug.	June–Aug.	Precipitation Gradient (inches/mile)
A	—	1.8	1.2	3.8	6.8	
B	2.8	2.8	1.5	5.4	9.7	
C	3.9	3.4	1.4	5.3	10.1	
D	4.3	1.4	0.7	3.5	5.6	
E	4.9	0.9	0.2	3.4	4.5	

EXERCISE 46

Compute in Table 7.6 the horizontal precipitation gradient between A and the other stations (difference in summer rainfall per mile). To which station is this gradient largest?

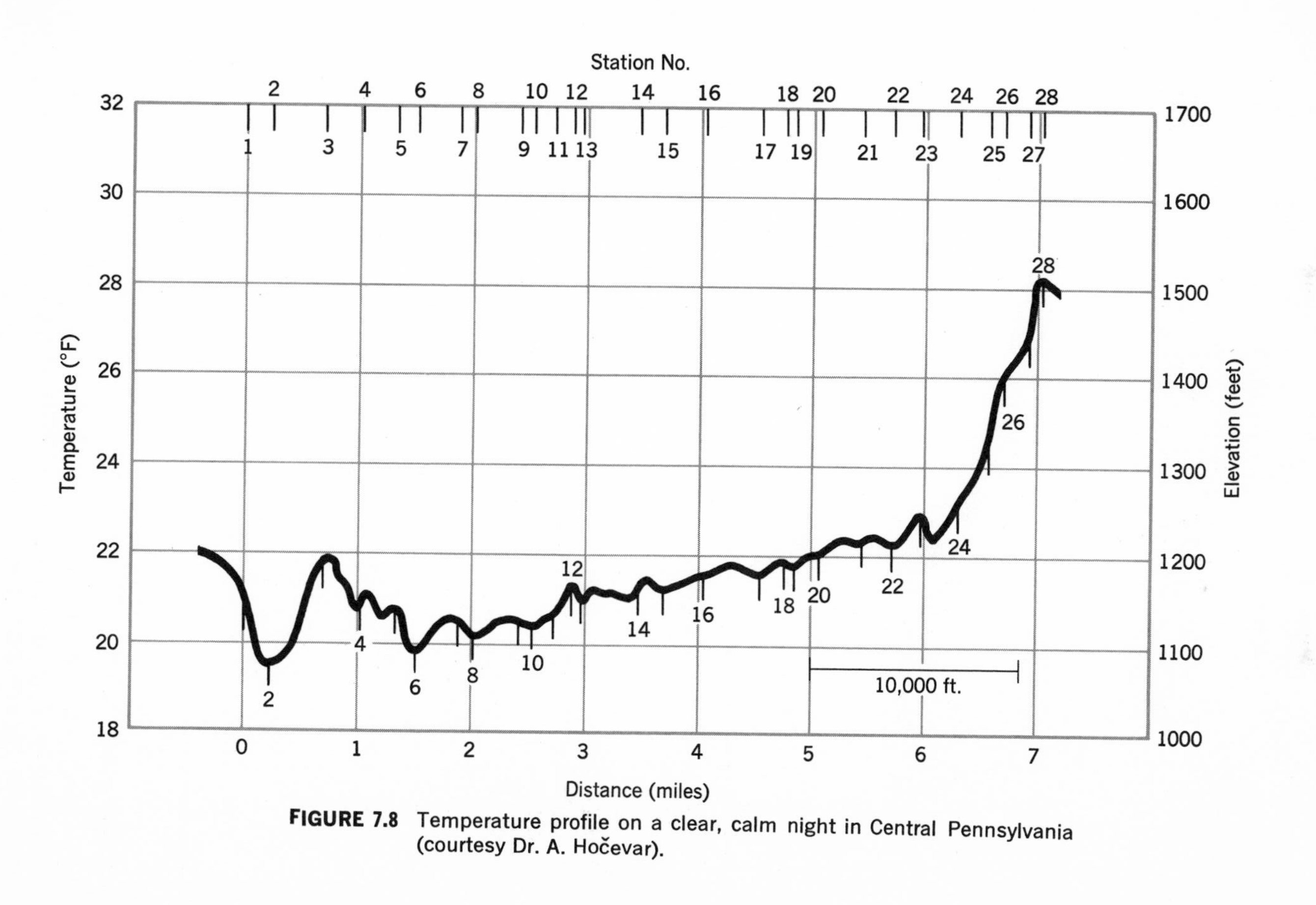

FIGURE 7.8 Temperature profile on a clear, calm night in Central Pennsylvania (courtesy Dr. A. Hočevar).

The distribution of rainfall in the Nittany Valley is essentially due to a combination of the effect of topographic contours and of preferred thunderstorm tracks across the valley. This is evident by the fact that Stations B and C have consistently higher, Stations D and E consistently lower rainfall than Station A.

To what extent even minor differences in elevation can cause large temperature variations over relatively small distances can be seen from temperature profiles measured in Central Pennsylvania; the measurements, made on 17 April 1968 between 0444 and 0534 EST, are reproduced in Table 7.7.

TABLE 7.7
Temperature Variations on a Clear Night in Central Pennsylvania (Courtesy Dr. A. Hočevar, University of Ljubljana, Yugoslavia)

Sta. No.	Temp. (°F)	Height (ft)	Sta. No.	Temp. (°F)	Height (ft)	Sta. No.	Temp. (°F)	Height (ft)	Sta. No.	Temp. (°F)	Height (ft)
1	27.0	1160	8	24.0	1110	15	26.0	1160	22	26.0	1215
2	26.5	1080	9	24.0	1130	16	24.5	1180	23	26.7	1255
3	29.7	1200	10	24.0	1125	17	22.5	1175	24	27.5	1260
4	24.0	1145	11	23.0	1135	18	25.0	1195	25	31.5	1340
5	25.2	1140	12	25.0	1165	19	25.0	1190	26	32.7	1400
6	23.0	1090	13	22.5	1150	20	25.0	1205	27	30.8	1440
7	26.0	1130	14	25.2	1175	21	24.7	1215	28	31.7	1510

EXERCISE 47

Plot the data on the graph of Figure 7.8. Note that the contour profile has a greatly exaggerated height scale, 100 ft of which correspond to 4000 ft on the horizontal distance scale.

a) Considering the fact that apple blossoms will be killed by a temperature of 28°F or less, at what elevation would an orchard have been completely safe from frost damage?

b) What is the explanation for the relatively high temperature at Station No. 3?

c) What is the range of temperatures observed on that night?

d) What is your general conclusion about the variation of minimum temperatures over relatively small areas on clear, calm nights?

7.5 URBAN EFFECTS

It is a well-known fact that many species of the animal world can and do modify their immediate climatic environment on a small scale, for example, warming and ventilation of hives by bee populations, to produce more favorable conditions. In this respect, man is no different, except that he is capable of more drastic modifications on a larger scale; while his activity is usually intended to be constructive, he often inadvertently creates conditions that are detrimental, such as air and water pollution.

In Table 7.8 the averages of the amount of dust in the air clearly show the effects of human habitation on air pollution, as measured by the U. S. Public Health Service.

TABLE 7.8
Average Dust Concentrations for U. S. Communities between 1953 and 1967 (After D. M. Anderson, J. Lieben, and V. H. Sussman)

	CONCENTRATIONS ($\mu g/m^3$)
Rural Areas	40
Suburban Areas	69
Cities of	
less than 0.7 million	110
0.7–1.0 million	150
larger than 1.0 million	190

EXERCISE 48

a) List some of the natural dust sources.

b) List some of the man-made dust sources.

Significant changes of the natural landscape are produced by the spread of cities and traffic arteries; not only is the topography altered by high-rise buildings, but the thermal nature of the surface is changed when vegetation, swamps, and so on, are replaced by concrete, stone, metal, and glass. The attendant effects on the local climate are negligible for small communities, but probably assume substantial significance in metropolitan areas.

The following factors must be considered: As compared to the undisturbed surrounding countryside, the complex vertical structure of a city causes turbulence and vertical mixing in the air layer near the surface. The city streets channel the wind, and tall buildings decrease its speed. The impermeable city surface makes precipitation ineffectual; only in park areas can percolation into the ground and evapotranspiration into the air take place. In the absence of such parks, flash flooding can become a real menace in summer thunderstorms.

Incoming solar radiation at street level is reduced, more because of the shading by buildings than because of depletion by air pollution. The net outgoing radiation is also reduced by back-radiation from polluted air layers and from buildings; in addition, the heating of buildings in winter, air conditioning in summer, motor traffic, and industrial activities are substantial heat sources that tend not only to warm the air, but also to increase convective activity with attendant cloud formation and precipitation.

The quantitative appraisal of city-induced climatic changes is not easy; while observational data from stations within a city and at nearby airports are available, comparisons are not clear-cut for the following reasons: Airports, although usually surrounded by natural landscape or smaller surburban communities, have large paved surface areas; they also may be located downwind from the metropolitan area. Weather stations in cities are usually located on the roofs of tall buildings, on hills, or in park areas, and therefore will not furnish wind and temperature information representative of the inhabited levels of the city. In particular, roof exposures of thermometers and rain gages are almost as questionable as they are common; exhausts from nearby buildings may affect the measured temperature, and the turbulent airflow over the roof may significantly alter the precipitation catch of the gages.

Nonetheless, comparison of the records collected at stations within a large city and at an out-of-town airport provides a suggestion that cities are indeed warmer and wetter than the surrounding countryside. The records at Philadelphia Airport, Pennsylvania, a few miles southwest of the city, and at the downtown campus of the Philadelphia Drexel Institute of Technology illustrate the situation rather well. Elevations are virtually the same at these locations, and the data of Table 7.9, each taken for the period 1951–1960, show the systematic differences.

EXERCISE 49

a) Is the greater warmth in the city more prominent at night or in daytime? In winter or in summer? State your reasoning.

b) Does it rain more frequently or more intensely in the city or both?

The mean temperature differences between city and suburb are rather small and may not be entirely traceable to differences in surface type and configuration. Considerable evidence has been amassed in recent years, supporting the existence of an "urban heat island"; however, many large cities are seaports so that continental–maritime contrasts may be distorting the resulting temperature distributions. What is more,

TABLE 7.9
Ten-Year Averages for Philadelphia Airport and Philadelphia Drexel Institute of Technology in the City

STATION	AIRPORT	DREXEL I. T.
ELEVATION (ft)	13	30
	°F	°F
MEAN ANNUAL TEMP.	55.0	56.5
MEAN ANNUAL MAX. TEMP.	64.0	64.7
MEAN JUNE MAX. TEMP.	82.1	82.8
MEAN DEC. MAX. TEMP.	43.5	44.1
MEAN ANNUAL MIN. TEMP.	45.9	48.3
MEAN JUNE MIN. TEMP.	61.7	63.9
MEAN DEC. MIN. TEMP.	28.3	31.2
MEAN ANNUAL PRECIP. (inches)	40.85	43.62
MEAN ANNUAL NO. OF DAYS WITH > 0.50 INCH PRECIP.	31	31

TABLE 7.10
Average Annual Number of Days with ≥ 90°F and ≤ 32°F at Various City (C) and Airport (A) Stations (1951–1960)

STATION		≥90°F	≤32°F	STATION		≥90°F	≤32°F
Los Angeles, Calif.	C	23	0	Knoxville, Tenn.	C	67	63
	A	3	0		A	51	73
Chicago, Ill.	C	18	103	Houston, Tex.	C	95	9
	A	27	123		A	94	9
Baltimore, Md.	C	35	62	Seattle, Wash.	C	2	16
	A	33	96		A	3	36
Pittsburgh, Pa.	C	19	96	Philadelphia, Pa.	C	32	73
	A	9	124		A	25	89
Washington, D.C.	C	39	68	Oklahoma City, Okla.	C	89	79
	A	33	72		A	76	81
Madison, Wisc.	C	9	143	Augusta, Ga.	C	96	34
	A	18	159		A	88	54
Indianapolis, Ind.	C	36	106	Des Moines, Iowa	C	31	137
	A	23	124		A	30	143

the experience of "feeling warmer" in the city is not wholly attributable to the air temperature, but rather to the "cooling power" of the air which is a combination of the effects of temperature, radiation, wind speed, and humidity on the rate at which the human body loses heat. In the city, decreased ventilation and additional radiation from buildings diminish this heat loss and produce a sensation of greater warmth of the

environment. It does seem safe to say, however, that the establishment and growth of an urban complex produces a changed climatic environment.

The discussion above well illustrates what is often the case in climatology, namely that mean values of climatological data can be difficult to interpret. In place of mean temperatures, the number of days on which the temperature exceeds or equals 90°F (or some other suitable limit) and the number of days on which the temperature is lower than, or equal to, 32°F (or some other limit) give a useful measure of temperature climate. In appraising the effects of cities, it might be expected that city stations should have a larger number of hot days and a smaller number of cold days than airports. To evaluate this hypothesis, Table 7.10 contains these data for 14 locations in the United States.

EXERCISE 50

a) Do urban locations tend to be warmer?

b) Is the high or low temperature limit more revealing of the urban effect?

c) How do you account for the anomalous situation with respect to high temperatures at both Chicago and Madison?

The effects of urban areas on the climate can be summarized as follows: While the alteration of the composition of air is often severe to the point of constituting a health menace, the effects on temperature and precipitation are rather subtle. The dust-loading of the air produces a moderate reduction of solar energy, but the interception of this energy by buildings somewhat compensates the loss. With respect to outgoing terrestrial radiation, these same factors also reduce it. The higher dust and smoke concentration in cities also provides an excellent source of condensation nuclei, so that cloudiness and reduced visibility are increased as well. Furthermore, temperatures are increased, but more so at night and in winter, and precipitation tends to be greater over the city. Nevertheless, the variation of temperature within a city, or even over a large farm, may be considerably greater than the differences between city and surrounding area. An example of this is given in Figure 7.9 in which the lines of equal temperature measured two meters above the ground in San Francisco, California, at 11 p.m. on 4 April 1952 are reproduced; the business district with its skyscrapers (right side in center of figure) is a good illustration of the urban heat island. Temperatures there are in the upper 60s, whereas the undeveloped area of the Golden Gate Park (upper left side of figure) is almost 20 degrees cooler although this area is still within the city limits.

FIGURE 7.9 Temperature distribution in San Francisco, California, at 2320 PST on 4 April 1952, showing a range of temperature at 2 meters from 49°F at Golden Gate Park (upper left) to 68°F in the business district (center right) (after Fowler S. Duckworth and James S. Sandberg).

Similarly, the observed precipitation may differ greatly in various parts of a city as can be demonstrated by data from rain gages within the metropolitan area of Seattle, Washington; four stations along a straight line from north–northeast to south–southwest with a maximum distance of 13 miles between the first and the last station are listed in Table 7.11. (See Figure 7.10.) In this case, the differences in precipitation are not easily ascribed to any specific cause; undoubtedly, the exposure of the rain gages plays some role, but more so the relative position of the stations in the rainshadow produced by the Olympic Mountain range, as well as the storm-channeling effect of the Strait of Juan de Fuca and the Puget Sound.

TABLE 7.11

Average Annual Precipitation at Various Stations in Seattle, Wash., for the Period 1931–1955

Station	Height (feet)	Precipitation (inches)
Bothell 2N	100	39.3
NAS	21	33.7
University of Washington	60	34.8
Federal Office Building	83	33.9

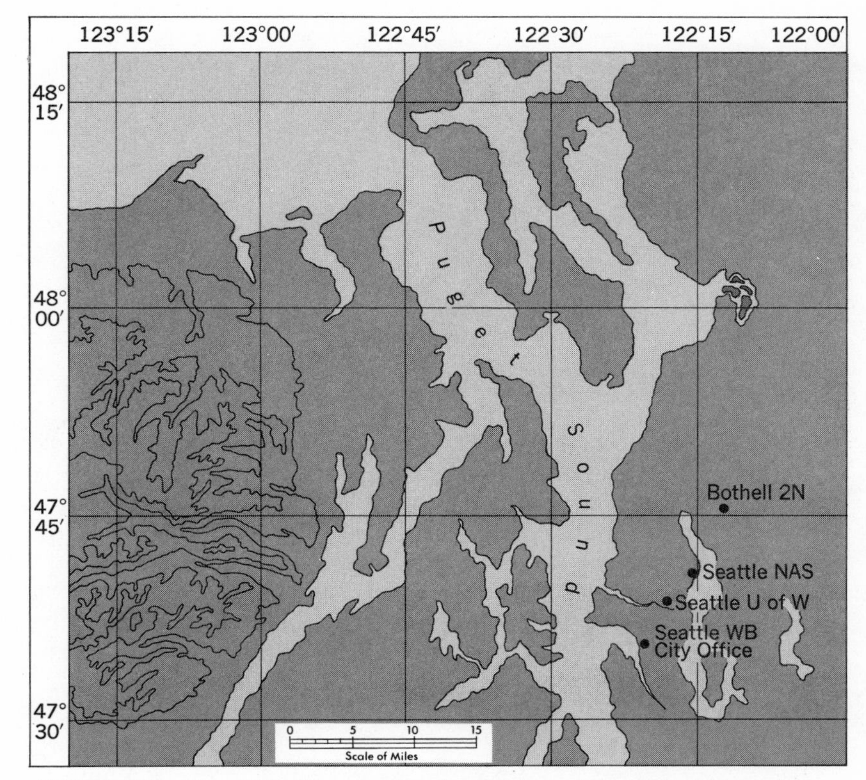

FIGURE 7.10 Map of Seattle area showing location of climatological stations.

climatic data sources

The following is a partial list of Government publications of climatic data available from the Superintendent of Documents, U. S. Government Printing Office, Washington, D. C. 20402.

Climatological Data — monthly records of temperature and precipitation with other elements for selected stations (by states). Annual summary includes normal values. $0.20 each.

Climatological Data, National Summary — more detailed monthly data from major stations together with narrative description of weather, maps, and charts. $0.20 each. Annual summary includes normals and extremes. $0.40 each.

Climatic Summary of the United States — Supplement for 1951–1960. — Temperature and precipitation data for the decade together with averages for longer periods at many stations (by states). $0.20 to $0.70 each.

Summary of Hourly Observations (1951–1960) — detailed hourly data for selected cities. $0.10 each.

World Weather Records — Temperature, precipitation, and pressure for numerous stations around the world (by continents). Six volumes from $2.50 to $4.00 each.

Storm Data — Narratives and data on severe storms in the United States (monthly). $0.15 each.

Climates of the States — data for each state including some charts (by states). $0.10 to $0.25 each.

Other information can be obtained from the National Weather Records Center, Asheville, North Carolina, 28801.

list of suggested reading

Barger, G. L. (Ed.). *Climatology at Work*. Asheville, N. C.: U. S. Department of Commerce, 1960, 109 pp.

Barry, R. G. and R. J. Chorley. *Atmosphere, Weather and Climate*. London: Methuen, 1968, 319 pp.

Conrad, V. and L. W. Pollak. *Methods in Climatology*. Cambridge, Mass.: Harvard University Press, 1950, 2nd edition, 459 pp.

Day, John A. *The Science of Weather*. Reading, Mass.: Addison-Wesley, 1966, 214 pp.

Dunn, G. E. and B. I. Miller. *Atlantic Hurricanes*. Baton Rouge, La.: Louisiana State University Press, 1964, 2nd edition, 377 pp.

Flora, S. D. *Tornadoes of the United States*. Norman, Okla.: University of Oklahoma Press, 1954, 221 pp.

Geiger, R. *The Climate Near the Ground*. Cambridge, Mass.: Harvard University Press, 1965, 2nd edition, 611 pp.

Griffiths, John F. *Applied Climatology*. London: Oxford University Press, 1966, 118 pp.

Haurwitz, B. and J. M. Austin. *Climatology*. New York: McGraw-Hill, 1944, 410 pp.

Landsberg, Helmut. *Physical Climatology*. DuBois, Pa.: Gray Printing Co., 1958, 2nd edition, 446 pp.

Panofsky, Hans A. and Glenn W. Brier. *Some Applications of Statistics to Meteorology*. University Park, Pa.: The Pennsylvania State University, 1958, 224 pp.

Petterssen, Sverre. *Introduction to Meteorology*. New York: McGraw-Hill, 1958, 2nd edition, 327 pp.

Rumney, George R. *Climatology and the World's Climates*. New York: The Macmillan Company, 1968, 656 pp.

Scorer, R. S. *Weather*. London: Phoenix House Ltd., 1959, 63 pp.

Sellers, W. D. *Physical Climatology*. Chicago, Ill.: University of Chicago Press, 1965, 272 pp.

Spar, Jerome. *Earth, Sea, and Air*. Reading, Mass.: Addison-Wesley Publishing Company, 1962, 152 pp.

Trewartha, G. T. *An Introduction to Climate*. New York: McGraw-Hill, 1968, 4th edition, 402 pp.

appendix I experiments

In several sections of the text, certain statements lend themselves to demonstration or testing by means of simple experiments. Three samples that can be performed with materials commonly found around the house and with inexpensive thermometers are given below.

EXPERIMENT 1: THE SUN AND THE SEASONS

From Figure I.1 the construction of the demonstration model is evident; a circular disk of 8-inch diameter represents the earth and is fastened with a paper fastener through its center to a 10 × 12-inch cardboard on which parallel sun's rays and the outer limit of the atmosphere (as a ring $9\frac{1}{2}$-inch diameter) are drawn. (For classroom demonstration, this model can be made much larger.) From the geometry of the situation, the sun's angular elevation h above the horizon at noon for which the model is valid can be determined for the northern hemisphere from $h = 90° - L + D$, where L is the latitude (positive for the northern, negative for the southern hemisphere) and D is the sun's declination, that is, the latitude at which the sun is in the zenith at noon. The declination varies from $+23\frac{1}{2}$ degrees to zero to $-23\frac{1}{2}$ degrees according to the date, and the values are printed every year in the ephemeres of sun for navigation purposes. An abbreviated version is reproduced in Table I.1. The above formula for the southern hemisphere reads $h = 90° + L - D$.

A. The spring and autumn equinoxes are represented by the model if the equator is lined up with the heavy center ray of the sun so that the earth's axis is perpendicular to the sun's rays.

a) What is the sun's noon elevation angle at latitudes 0, 30, 45, 60, and 90 degrees north and south?
b) How does the pathlength of the sun's rays through the atmosphere vary with latitude and sun's elevation?
c) What can you say about the insolation at low, middle, and high latitudes?

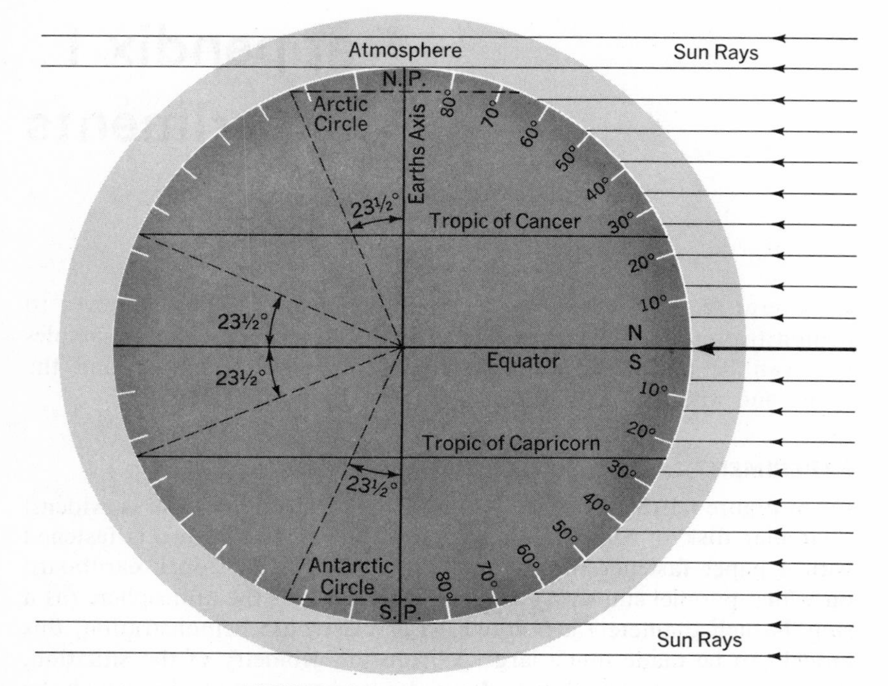

FIGURE I.1 The sun and the seasons.

TABLE I.1
Abbreviated Variation of Sun's Declination during the Year

DATE		D	DATE		D	DATE		D	DATE		D
Jan.	1	−23.1	Apr.	1	+4.1	July	1	+23.2	Oct.	1	−2.7
	11	−22.0		11	+7.9		11	+22.3		11	−6.6
	21	−20.2		21	+11.4		21	+20.7		21	−10.3
Feb.	1	−17.4	May	1	+14.7	Aug.	1	+18.3	Nov.	1	−14.0
	11	−14.4		11	+17.6		11	+15.6		11	−17.1
	21	−11.0		21	+19.9		21	+12.5		21	−19.7
Mar.	1	−8.0	June	1	+21.9	Sep.	1	+8.7	Dec.	1	−21.6
	11	−4.2		11	+23.0		11	+5.0		11	−22.9
	21	−0.2		21	+23.4		21	+1.2		21	−23.4

B. The earth's disk is turned to its position during the summer solstice by lining up the Tropic of Cancer (23½°N latitude) with the center ray.

a) How high is the noon sun at the equator?
b) Is the sun in the northern or southern sky there?
c) How high is the sun at the north and south pole, respectively?
d) At which latitude is the sun in the zenith?
e) What are the daylengths north of the arctic and south of the antarctic circles, respectively?
f) What can you say about the insolation at 45°N and 45°S latitudes as compared to that during the equinoxes?

C. Finally the earth's disk is turned to its position during the winter solstice by lining up the Tropic of Capricorn (23½°S latitude) with the center ray. Answer the same questions as in section B **a)** to **f)** for this time of the year.

EXPERIMENT 2: ABSORPTION OF RADIANT ENERGY BY SOIL

Take some fine, sifted soil and spread it evenly on a glass plate or sheet of transparent plastic held over a bright light source, such as a 200-watt bulb or a flood light. The glass-plate size does not have to be large, a few square inches are sufficient.

a) How thick a soil layer is needed to obscure the light completely? This layer, then, absorbs all the radiant energy.
b) Do you think you could similarly find the thickness of a layer of water that would also absorb all that energy?
c) How would you go about setting up a suitable experiment to test this? Is it feasible?

EXPERIMENT 3: DIFFERENCES IN THERMAL RESPONSES OF VARIOUS SURFACES

For this experiment, it is best to use a liquid-in-glass thermometer that has no backing behind the bulb, such as an ordinary laboratory thermometer with graduation to at least 120°F. An inexpensive thermometer is good enough, if the backing behind the bulb can be removed without damage to the thermometer.

a) On a sunny day with light winds or calm, first measure the air temperature in the shade by moving the thermometer briskly through the air, with arm outstretched. In this manner, ventilate the thermometer for two to three minutes, then read it.

b) Measure the temperature of bare, loose soil in sunshine by inserting the bulb so that it is just barely covered. Read the thermometer when the temperature stops rising or rises only very slowly.

c) Then, after cleaning the dirt off the bulb, measure the air temperature again in the shade.

d) Then measure the temperature of the grass by bending a bundle of grass blades down to the ground, laying the thermometer bulb on top and covering it with more grass. In doing so, do not touch the bulb or the portions of grass that will be in contact with the bulb. Read the thermometer when it has stopped changing noticeably.

e) Repeat the air temperature measurement in the shade.

f) If available, the temperature of a water surface (puddle, pond, creek, and so on) should be similarly measured by insertion of the bulb just under the surface; otherwise the bare soil can be watered and its temperature measured as before.

g) Average the air temperatures and compare with those of the various surfaces and explain the cause of the differences.

appendix II climatic instruments and observations

In the following, various possibilities for climatological observations that require only inexpensive thermometers are described. Where funds are available, a commercial sling psychrometer, a thermometer shelter with maximum and minimum thermometers, a rain gage, a wind vane, and anemometer are recommended.

It is important that observations are made systematically, that is, every day at the same time and place for an uninterrupted period of sufficient length, to have data that can be used for determining relationships among climatic elements. If this is not possible, it is better to use observations published by the U. S. Weather Bureau. (See List of Government Publications, page 133.)

1. DRY-BULB AND WET-BULB TEMPERATURE MEASUREMENTS

An instrument that consists of a dry-bulb and a wet-bulb thermometer mounted together is called a psychrometer. The rate of evaporation of water from the wet bulb and therefore its cooling depends on the relative humidity, being greater when the relative humidity is low. The construction of a psychrometer should be clear from Figure II.1. It is important to use a fine saw blade to cut away the lower end of the backings of two identical thermometers, so that the bulbs and about $\frac{1}{4}$ inch of the capillaries above the bulbs are free. Be sure not to disturb the mountings of the capillaries on the scales.

For outdoor measurements select an open place in the shade; wet the wick of the wet bulb with clean water. Then hold the sling psychrometer at arm's length and sling the thermometers around the handle three times per second; after a minute or so quickly read (without touching

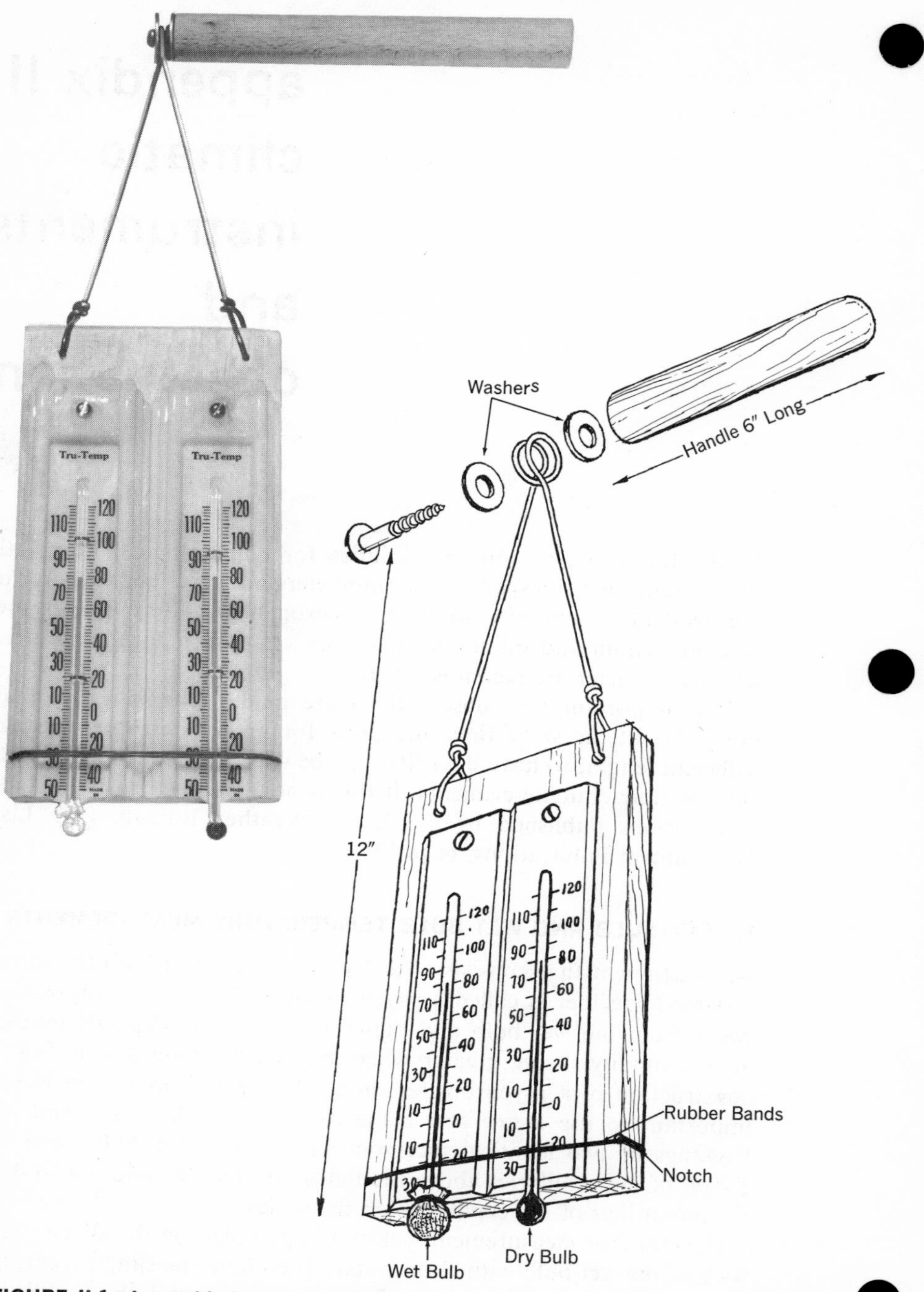

FIGURE II.1 Assembled psychrometer (left) and construction of psychrometer. (right).

the thermometers) first the wet-bulb, then the dry-bulb thermometer to the nearest $\frac{1}{2}$ degree, if possible. Repeat the procedure until two successive readings show no more temperature changes.

Psychrometric tables can be found in many elementary textbooks on meteorology or can be purchased from the Superintendent of Documents, Government Printing Office, Washington, D. C., for $0.25. With such tables, the relative humidity and the dew point can be determined from dry- and wet-bulb temperatures. Or the relative humidity (RH) can be determined with the rule of thumb:

$$\mathrm{RH} = 100 - \frac{300(T - W)}{T}$$

where T is the dry-bulb, W the wet-bulb reading; thus, if $T = 70°\mathrm{F}$ and $W = 63°\mathrm{F}$, $\mathrm{RH} = 100 - 300(70 - 63)/70 = 100 - 300 \times 7/70 = 100 - 30 = 70$ percent.

Always make sure that the wick is thoroughly wet; in dry weather the wick may dry out during ventilation. In such a case, it helps to use very cold water.

Materials: Two identical thermometers;
wooden board $4 \times 8 \times \frac{3}{8}$ inch, approximately;
stiff wire 15 inches long, $\frac{1}{16}$-inch diameter;
dowel rod 6 inches long, $\frac{3}{4}$-inch diameter;
three screws;
two washers;
thin, white cotton material one-inch square for wick and cotton thread for fastening it to bulb;
two strong rubber bands.

2. RAIN MEASUREMENTS

The amount of rainfall is measured as the depth of water on level, impermeable ground. Any cylindrical can may be used to catch precipitation; it is desirable to determine the depth of water in a can with an accuracy of 1/100 inch, so that it is necessary to magnify the depth of the rain catch. In addition, evaporation of water prior to measurement must be minimized; this is achieved by placing a funnel over the can.

The rain gage is constructed as illustrated in Figure II.2. Two large tin cans, from which the tops have been removed as well as the bottom of one of them, are soldered (with hot or cold solder) together. A disk that fits loosely inside the can is made of wood or stiff foam plastic, in the center of which a hole is made; the funnel neck is glued into it.

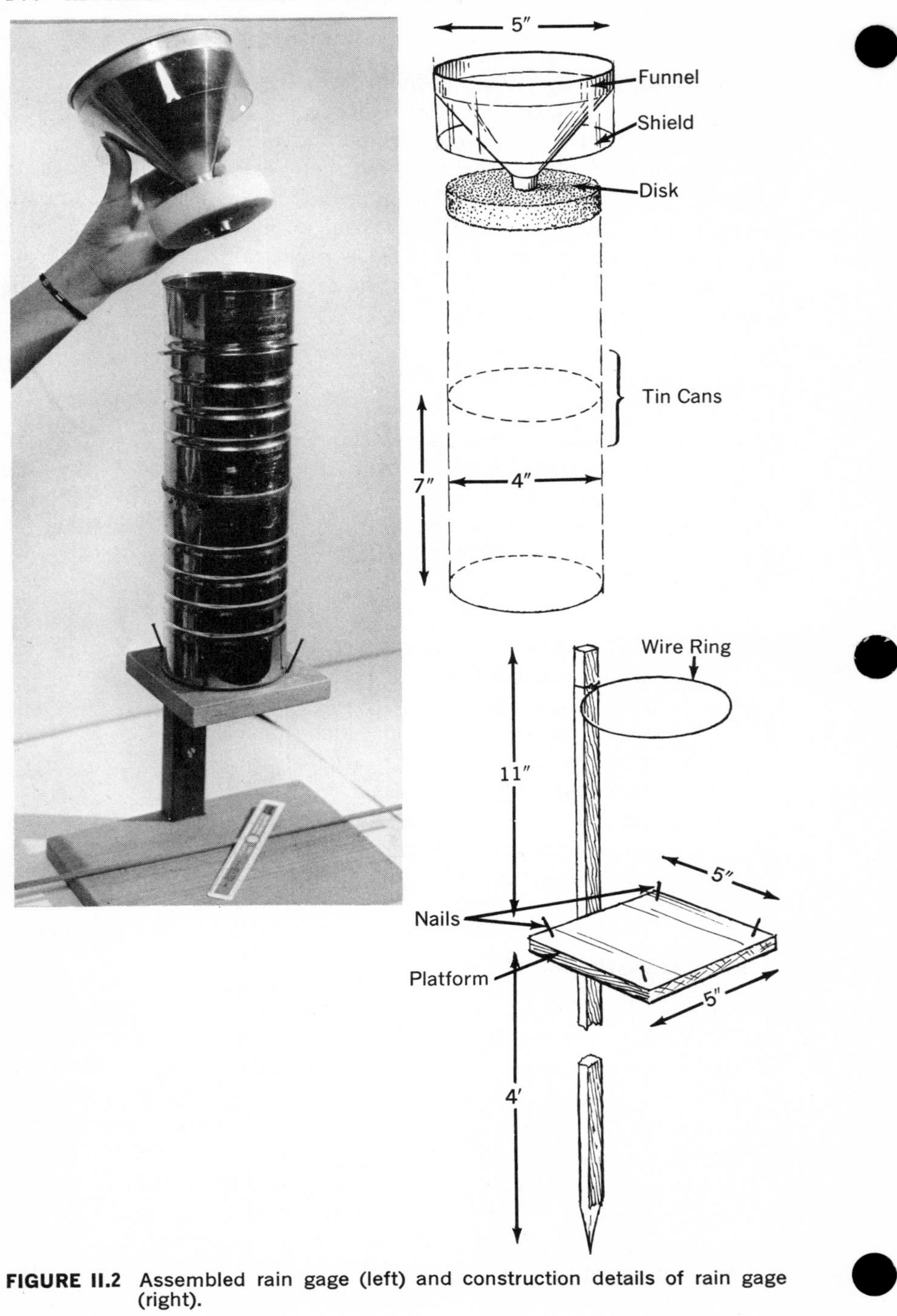

FIGURE II.2 Assembled rain gage (left) and construction details of rain gage (right).

A sheet of tin-can metal or plastic 2 × 15 inches is cemented around the funnel rim to prevent driving rain from getting into the can without going through the funnel. The finished gage is mounted on a post as suggested in Figure II.2, and should be positioned so that it is as far away from any objects as these objects are high.

For measurement of rainfall, the contents of the can are poured into another cylindrical can of much smaller diameter, preferably not more than 2 inches. Several such small tin cans can be soldered together in the same manner as the rain gage itself. With a very slender redwood stick, which shows watermarks well, we measure the height of the water column in the measuring tube. The depth of precipitation, P, is computed from the relationship

$$P = H\left(\frac{d}{D}\right)^2,$$

where H is the height of the watermark, d the inner diameter of the measuring tube, and D the inner diameter of the funnel top. For example, let D be $5\frac{3}{8}$ inches, $d = 1\frac{1}{8}$ inches; thus the factor

$$\left(\frac{d}{D}\right)^2 = \left(\frac{9/8}{43/8}\right)^2 = (0.21)^2 = 0.044.$$

So if the length of the watermark was $H = 1\frac{3}{16}$, the precipitation amount was $P = 1.19 \times 0.044 = 0.05$ inch.

When precipitation falls in form of snow, the funnel is removed. The snow caught in the can is melted prior to transfer into the measuring tube. This is best done by adding a known depth of warm water and subtracting that amount from the measured value.

Materials: Two large tin cans, at least 7 inches high and 4-inch diameter;
one metal funnel about 5-inch diameter;
wooden post $1 \times 1\frac{1}{2}$ inches cross section, 3 to 4 ft long;
$5 \times 5 \times \frac{3}{4}$ inches wooden board;
$1 \times 4 \times 4$ inches wood or stiff foam plastic;
strip of sheet metal or plastic 2 × 15 inches;
18 inches stiff wire;
redwood stick $\frac{1}{8} \times \frac{3}{16} \times 12$ inches;
two or three tin cans of small diameter;
nails;
glue;
liquid (or hot) solder.

3. VISIBILITY OBSERVATIONS

"Visibility" is the farthest horizontal distance, in miles and/or fractions of miles, at which prominent dark objects can be seen against the sky. The transparency of the air is a characteristic of air masses. For climatological purposes, it is sufficient to estimate visibility according to a simple qualitative scale:

1) "poor" for hazy or foggy conditions;
2) "fair" for average conditions;
3) "good" for clear-air conditions;
4) "excellent" for extremely clear air.

4. CLOUDINESS OBSERVATIONS

The radiation climate in a given region is, to a large extent, determined by the cloudiness. The relative amount of sky covered by clouds is most easily estimated according to the aviation weather reporting code:

CLEAR, represented by the symbol ○, denotes that no clouds are present.
SCATTERED, ⦶, means clouds cover only half of the sky or less.
BROKEN, ⦶, means clouds cover more than half of the sky, but some blue area is still visible.
OVERCAST, ⊕, is the case of the sky that shows no blue area.
OBSCURED, ⊗, is the case of the sky not being visible on account of dense fog, smoke, and so on.

5. BLUENESS OF SKY

This characteristic in the atmosphere is related to humidity, air pollution, visibility, and so on. It can be qualitatively estimated at the spot of deepest blue in the sky in the following categories:

No. 1 — "pale blue" for a milky blue sky;
No. 3 — "medium blue" for an average blue sky;
No. 5 — "deep blue" for a very deep, pure blue sky.

If the sky color is estimated between the three categories given, the even numbers are recorded; thus No. 4 is used for a sky that is darker blue than average but not quite "very deep" blue. No. 0 is used for an overcast sky.

6. WIND OBSERVATIONS

Observations of wind are best made in the middle of an open field with short grass or bare ground. The wind direction, that is, the direction from which the wind blows, can be ascertained by "feeling" the wind with one's face, by observing smoke drift, flags, and so on. The wind speed can be estimated according to the Beaufort scale in Table II.1.

Instrumental observations of wind are made with a vane and anemometer. (See Figure II.3.) A wind vane can be constructed with the materials from List a) at the end of Appendix II according to the dimensions given in Figure II.3. Note that before drilling the hole in the vane shaft, the assembled vane is balanced on a knife edge to find the center of gravity. A nail that fits loosely through the shaft is then hammered into a dowel rod which serves as a handle. If a piece of soldering wire or other heavy material is wound around, or otherwise fastened to, the front end of the vane, the center of gravity shifts closer to the middle of the shaft.

For measuring wind speed, a cup anemometer can be built from three small aluminum funnels and the other materials in List b). The funnel spouts are cut away as shown in Figure II.3 and then closed off with liquid solder. The remaining construction and assembly can be gleaned

from the figure. Finally, one cup is painted flat black for easier counting of the revolutions.

For high wind speeds, when it is difficult to count the large number of revolutions per minute, another cup arrangement can be used, namely half-cylinders bent from $2\frac{1}{2} \times 5$ inches tin-can metal as shown in Figure II.3. This type of anemometer will respond only to fairly strong winds.

In order to calibrate the anemometers, they are mounted on a broom stick so that the instruments can be held out of a car window well above the car top. While the driver drives at constant speed along a straight stretch of road on a calm day, the observer counts the revolutions per minute; this is done for speeds of 5, 15, 25, and 40 mph, and a calibration graph is constructed to permit easy conversion of rpm into mph.

Materials: a) Sheet of tin-can metal, 4×15 inches
Sheet of tin-can metal, $4\frac{1}{2} \times 5$ inches
Strip of tin-can metal $\frac{1}{2}\times 4$ inches
One $\frac{1}{2}$-inch dowel rod 18 inches long
One stick of wood 24 inches long
Glue, nails, one glass bead
b) Three aluminum funnels $2\frac{1}{2}$ inches wide
Three $\frac{3}{16}$-inch dowel rods, 7 inches long
One metal (or plastic) pencil cap

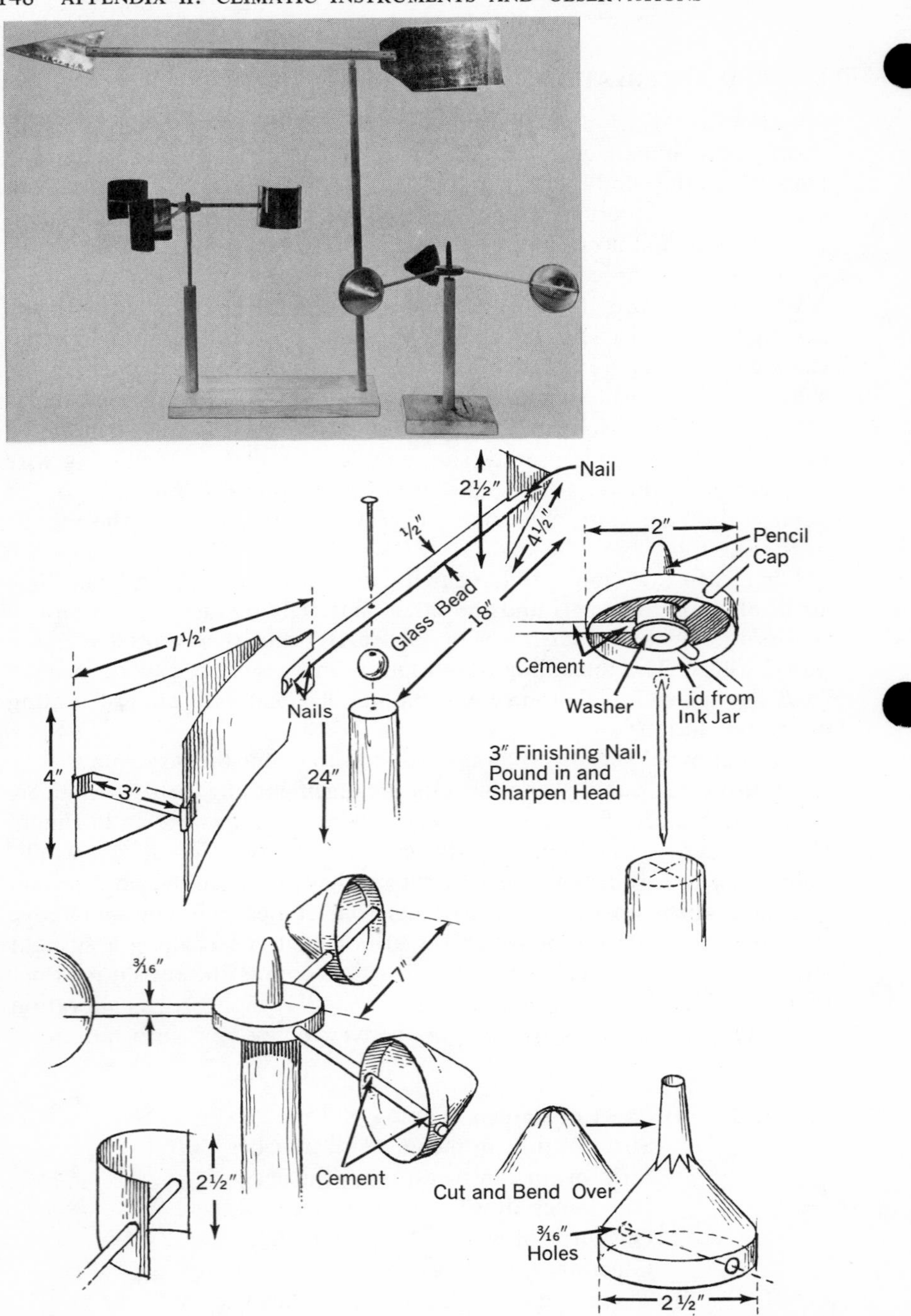

FIGURE II.3 Assembled wind vane and anemometers (top); construction of wind vane and anemometers (bottom).

One metal lid $1\frac{3}{4}$ to 2 inch diameter (such as from ink bottle)
One 3-inch nail
One small metal washer, $\frac{3}{8}$-inch diameter
One stick of wood 24 inches long ($\frac{3}{4}$-inch dowel rod)
Glue, flat black paint

c) Three pieces of tin-can metal, $2\frac{1}{2} \times 5$ inches
Three $\frac{3}{16}$-inch dowel rods, 7 inches long
One metal (or plastic) pencil cap
One metal lid, $1\frac{3}{4}$-to 2-inch diameter
One small washer, $\frac{3}{8}$-inch diameter
Glue, flat black paint

TABLE II.1
Beaufort Wind Scale

No.	Description of Wind Effect	Equivalent in Knots
0	Smoke rises vertically	less than 1
1	Wind direction shown by smoke drift but not by wind vane	1–3
2	Wind felt on face; leaves rustle; vane moved by wind	4–6
3	Leaves and small twigs in constant motion; wind extends light flag	7–10
4	Raises dust and loose paper; small branches are moved	11–16
5	Small trees in leaf begin to sway; crested wavelets form on inland water	17–21
6	Large branches in motion; whistling heard in telegraph wires; umbrellas used with difficulty	22–27
7	Whole trees in motion; inconvenience felt in walking against wind	28–33
8	Breaks twigs off trees; generally impedes progress in walking against wind	34–40
9	Slight structural damage occurs; chimney pots and slate removed	41–47
10	Seldom experienced inland; trees uprooted; considerable damage occurs	48–55
11	Very rarely experienced; widespread damage	56–65
12	Hurricane force	above 65

appendix III use of climatic data

Making observations of climatic elements is a means toward an end, namely the utilization of the data for getting an insight into, or understanding of, interrelations among the observed elements, their variations with time and place, and so on. For these purposes, the data have to be suitably summarized and analyzed; conclusions must then be drawn from the analyses and tested as to their validity. A few simple methods are described here; for more sophisticated statistical techniques, consult texts by H. Landsberg or H. Panofsky and G. Brier (see List of Suggested Reading).

1. DEFINITIONS

The daily mean temperature is the sum of the maximum and minimum temperatures during a 24-hour period divided by 2. If the maximum and minimum thermometers are read in the early morning, the minimum temperature is for the day on which it has been read, the maximum temperature for the preceding day. The monthly mean temperatures are the sums of the daily maximum and minimum temperatures divided by twice the number of days in the month. The annual mean temperature is the sum of the mean monthly temperatures divided by 12. The daily range of temperature is the difference between the highest and lowest temperature observed for that day; the monthly range is the difference between the mean temperatures of the warmest and coldest days during the month, and the annual range is the difference between the mean temperatures of the warmest and coldest months during that year.

For precipitation, the sums are usually of interest, rather than the means. In addition, the number of days on which a climatic element occurred or on which a given element had values above or below certain limits, such as the number of days with rain exceeding 0.50 inch, and so

on, is often useful. This last method is particularly applicable to categorized data, such as cloudiness, for which one often determines the number of clear days per month (or year), that of scattered clouds, or any other category.

Categories can also be used for continuous numerical data such as temperature, which can be divided into class intervals, for example 5°F intervals, such as 30–34, 35–39, 40–44, 45–49°F; similarly, wind speeds can be categorized in classes of 0–2 mph, 3–5 mph, 6–10 mph, 11–20 mph, and larger than 20 mph. In other words, classes can have equal intervals such as in the temperature example, or unequal intervals as the wind speed, where certain categories predominate.

2. RELATIONSHIPS AMONG NUMERICAL VARIABLES

As an example, we may wish to investigate whether or not a relationship exists between wind direction and rainfall. We can ask: What are the wind directions with which precipitation is most likely to occur? To answer this question, we determine the number of hours (or days) during which north wind prevailed and the percent of these periods of north wind on which precipitation was observed; we do the same for the other wind directions and calm periods, separately. The results can then be graphically displayed as in Figure III.1, where we see that at University Park, Pennsylvania, easterly winds favor the occurrence of precipitation, whereas only $2\frac{1}{2}$ percent of the hours with northwest winds are accompanied by precipitation. This distribution is explained by easterly winds occuring when the polar front is a short distance to the south, whereas northwest winds are observed after the passage of cold fronts that bring cold, dry weather to the region.

TABLE III.1
Average Hourly Temperature and Relative Humidity for 2–4 June 1967 at University Park, Pennsylvania

Time (hr)	Temp. (°F)	RH (percent)	Time (hr)	Temp. (°F)	RH (percent)	Time (hr)	Temp. (°F)	RH (percent)
00	62	55	08	63	61	16	82	30
01	60	61	09	69	54	17	81	30
02	58	69	10	74	43	18	79	32
03	56	76	11	77	37	19	77	35
04	54	79	12	78	33	20	73	38
05	53	83	13	80	32	21	70	44
06	54	83	14	81	30	22	68	50
07	58	72	15	81	30	23	67	54

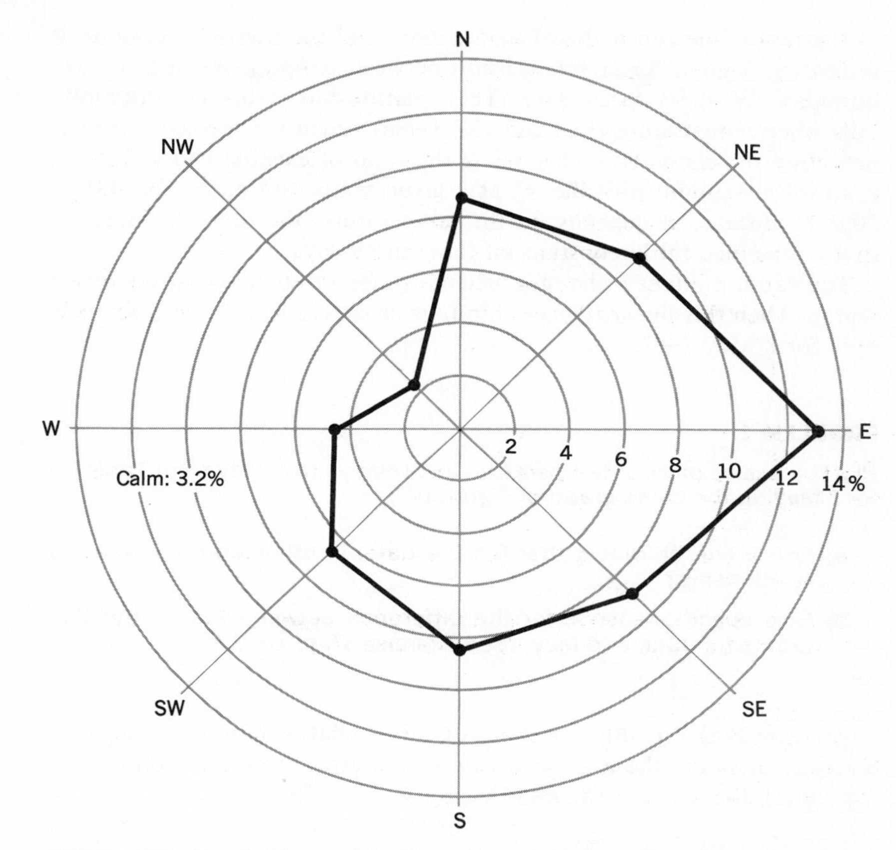

FIGURE III.1 Relative frequency of hourly precipitation with various wind directions at University Park, Pa.

Another method for determining relationships between two variables is the *scatter diagram*. As an example, the hourly averages of temperatures and relative humidity (RH) on three June days in 1967 at University Park are given in Table III.1. Each of the 24 pairs of data are plotted as one point on the graph of Figure III.2. The scales, increasing upward and to the right, respectively, are best chosen so that the ranges of values of each variable spans approximately equal lengths of the scales. Here, the temperature ranges from 53°F to 82°F, that is, roughly 30°, whereas the relative humidity covers the range from 30 to 83 percent, or about 50 percent, roughly twice as many percent as degrees. So the scales should be chosen so that the length of 10°F intervals is the same as the length of 20 percent relative humidity intervals.

A straight line can be fitted to the data, and the points lie close to it indicating a good linear relationship between temperature and relative humidity for these three days. This relationship is inverse (humidity falls when temperature rises and vice versa) which is expected from the definition of relative humidity being the ratio of existing vapor density, e, to the maximum possible, E, at a given temperature, that is, RH = $100e/E$. Because E depends on the temperature, the diagram indicates that e remained fairly constant on these three days.

The vapor content e changes, when a different air mass moves into a region. Then this linear relationship fails as is evident from the bihourly data for 2 May 1967.

PROBLEM 1

Plot the twelve pairs of temperature and relative humidity from Table 6.8 for 2 May on the same graph of Figure III.2.

a) Can a line be drawn that fits the data, in other words, is there a relationship?

b) Give specific reasons for the difference between the scatter diagrams for June and May. (See Exercise 37, p. 101.)

As regards the quality, or closeness, of a relationship as revealed by a scatter diagram, the less the data points scatter around a straight line (or curve), the better is the relationship.

3. RELATIONSHIPS AMONG CATEGORICAL VARIABLES

Another useful method for variates in categories can be demonstrated by considering the possible relationship between visibility and sky blueness, at University Park, Pennsylvania. Visibility was estimated in four categories (1 = poor; 2 = fair; 3 = good; 4 = excellent), blueness of sky in three categories which are somewhat different from those given in Appendix II (1 = pale blue; 2 = medium blue; 3 = deep blue). We construct a so-called *contingency table* with the visibility (V) categories as four columns, sky blueness (B) categories as three rows shown in Table III.2, and tally all simultaneous pairs of observations into the appropriate boxes. We then write the totals into the boxes and add them by columns and rows to get the column sums, t_c, and the row sums, t_r, as well as the total number of observed pairs, T. Because the majority of observations fall on a diagonal of the contingency table, a good relationship between the two variates is indicated: The blueness becomes

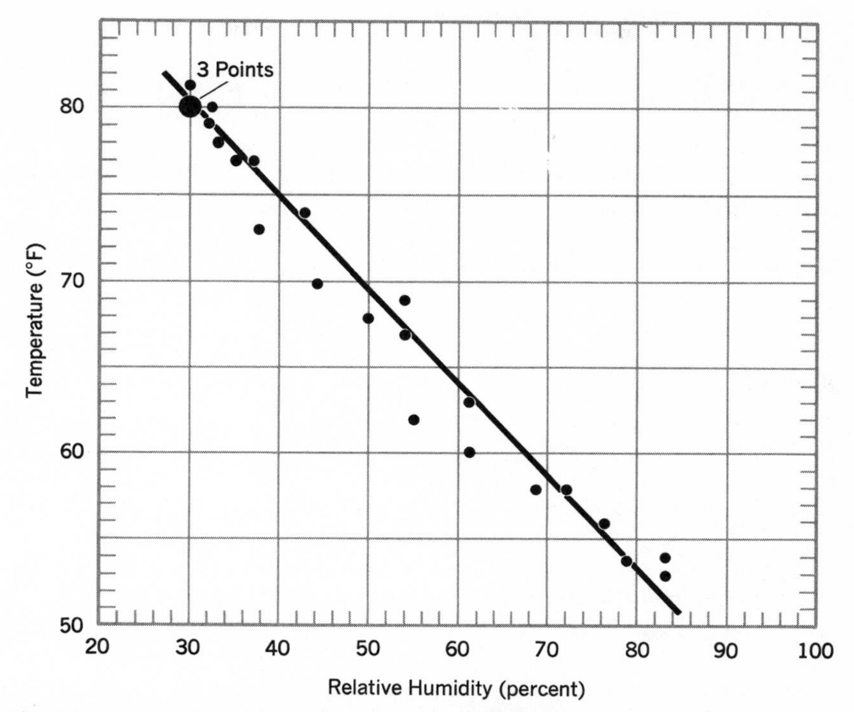

FIGURE III.2 Scatter diagram of the relationship between temperature and relative humidity on 2–4 June 1967 at University Park, Pa.

deeper as the visibility becomes greater. Of course, this does not imply that one of the elements is "caused" by the other; both may, and in fact do, have a common cause in a third factor: atmospheric suspensions. What is more, a random set of 130 pairs of observations could have produced the same kind of distribution *by chance*.

TABLE III.2

Contingency Table of Visibility versus Blueness of Sky at University Park, Pennsylvania

B \ V	1	2	3	4	t_r
1	𝍸 𝍸 𝍸 𝍸 𝍸 25	𝍸 /// 8	/// 3	0	36
2	𝍸 𝍸 10	𝍸 𝍸 𝍸 /// 18	𝍸 /// 8	/// 3	39
3	0	𝍸 //// 9	𝍸 𝍸 𝍸 𝍸 𝍸 25	𝍸 𝍸 𝍸 𝍸 / 21	55
t_c	35	35	36	24	$T = 130$

In order to test the *significance* of the observed distribution, we make another contingency table that has the same row and column sums, but the boxes will have "random" numbers. The theory of probability tells us what numbers we should put into the boxes: For each box we multiply the pertinent column sum, t_c, by the pertinent row sum, t_r, and divide by the total, T. For example, the upper left-hand box would get the number $35 \times 36/130 = 9$; the next box to the right: $35 \times 36/130 = 9$; the next $36 \times 36/130 = 10$; the next $24 \times 36/130 = 8$, and so on. The results of the chance contingency table are shown in Table III.3, using only whole numbers. Of the twelve boxes, we need compute only six, namely, two rows and three columns, because the row and column sums are the same as for the original contingency table. For example, the first column sum is 35, the first two chance numbers are computed at 9 and 11, therefore the third chance number must be $35 - (9 + 11) = 15$. Or the last row has the sum of 55 and the first three numbers are 15, 15, 15, whose sum is 45, so that the last number in that row must be $55 - 45 = 10$, and so on.

TABLE III.3
Chance Contingency Table Pertaining to Table III.2

B \ V	1	2	3	4	t_r
1	9	9	10	8	36
2	11	11	11	6	39
3	15	15	15	10	55
t_c	35	35	36	24	130

Note: This table should have no box with less than 5 in order to give reliable results.

The number of boxes for which we must compute the "chance" value by multiplying the row sums by the column sums and dividing by the total is called the *degrees of freedom* (DF), which in our example is 6. In the general case, the DF is the number of columns, c, minus one, times the number of rows, r, minus one, that is, DF $= (c - 1)(r - 1)$. Since for the above table $r = 3$, $c = 4$, DF $= (4 - 1)(3 - 1) = 6$.

Now for the significance test, called χ^2 test (chi square test):

$$\chi^2 = \frac{(O_1 - C_1)^2}{C_1} + \frac{(O_2 - C_2)^2}{C_2} + \cdots + \frac{(O_{12} - C_{12})^2}{C_{12}},$$

where O_1 is the observed number in the first box of Table III.2, C_1 the chance number in the first box of Table III.3; O_2 is the observed number in the second box of Table III.2, C_2 the chance number in the second box of Table III.3, and so on. Because the differences $O - C$ are squared, we can always subtract the smaller number from the larger number. The total number of terms in the equation is, of course, equal to the number of boxes. If we substitute the numbers from Tables III.2 and III.3, we obtain:

$$\chi^2 = \frac{(25 - 9)^2}{9} + \frac{(8 - 9)^2}{9} + \frac{(3 - 10)^2}{10} + \frac{(0 - 8)^2}{8} + \frac{(10 - 11)^2}{11}$$
$$+ \frac{(18 - 11)^2}{11} + \frac{(8 - 11)^2}{11} + \frac{(3 - 6)^2}{6} + \frac{(0 - 15)^2}{15}$$
$$+ \frac{(9 - 15)^2}{15} + \frac{(25 - 15)^2}{15} + \frac{(21 - 10)^2}{10}$$
$$= \frac{16^2}{9} + \frac{1^2}{9} + \frac{7^2}{10} + \frac{8^2}{8} + \frac{1^2}{11} + \frac{7^2}{11} + \frac{3^2}{11} + \frac{3^2}{6} + \frac{15^2}{15} + \frac{6^2}{15}$$
$$+ \frac{10^2}{15} + \frac{11^2}{10}$$

$$= \frac{256}{9} + \frac{1}{9} + \frac{49}{10} + \frac{64}{8} + \frac{1}{11} + \frac{49}{11} + \frac{9}{11} + \frac{9}{6} + \frac{225}{15} + \frac{36}{15} + \frac{100}{15} + \frac{121}{10}$$

$$= \frac{257}{9} + \frac{170}{10} + \frac{64}{8} + \frac{59}{11} + \frac{9}{6} + \frac{361}{15}$$

$$= 28.6 + 17.0 + 8.0 + 5.4 + 1.5 + 24.1 = 84.6$$

If all the observed values, O, were equal to the chance numbers, C, all the differences would vanish and χ^2 would be zero, indicating that there was no significance to the observed distribution, thus no relationship between visibility and sky blueness. But χ^2 is 84.6, which is very different from zero. Does this mean that there is, in fact, a relationship? Statistical significance tests never provide such guarantees but only confidence limits under which assertions can be made. The level of confidence must be chosen; then the degrees of freedom together with the value of χ^2 determine the acceptability of the assertion under the confidence limit chosen. The value of χ^2 must exceed a certain value for each set of conditions in order to indicate that the relationship is not a chance one. In Table III.4 these maximum χ^2 values are given for different degrees of freedom and two risks (= significance levels) of being wrong in 1 out of 100 cases (1 percent), and of being wrong in only 1 out of 1000 cases (0.1 percent). Comparing our value of 84.6 with the maximum values 16.8 for 1 percent risk and 22.5 for a 0.1 percent risk, we see that the probability of a real relationship between visibility and sky blueness to exist is considerably better than 99.9 percent.

This method can only be used for frequency numbers, not for physical quantities such as temperatures, wind speed, and so on. However, it is usually quite easy to convert physical quantities into categorical frequencies.

TABLE III.4*

Limiting χ^2 Values That Must Be Exceeded in Order to Assure Significance (Nonrandomness) at the Levels of 1.0 Percent and 0.1 Percent

DF	Significance 1 percent	Significance 0.1 percent	DF	Significance 1 percent	Significance 0.1 percent	DF	Significance 1 percent	Significance 0.1 percent
1	6.6	10.8	6	16.8	22.5	12	26.2	32.9
2	9.2	13.8	7	18.5	24.3	14	29.1	36.1
3	11.3	16.3	8	20.1	26.1	16	32.0	39.2
4	13.3	18.5	9	21.7	27.9	18	34.8	42.3
5	15.1	20.5	10	23.2	29.6	20	37.6	45.3

*This table is taken from Fisher & Yates' *Statistical Tables for Biological, Agricultural and Medical Research*, published by Oliver & Boyd Ltd., Edinburgh, and by permissions of the authors and publishers.

PROBLEM 2

In Table III.2 combine the visibility categories 1 and 2, as well as 3 and 4; also combine the blueness categories 1 and 2, leaving 3 separate, so that a new contingency table results with only four boxes. Then compute a new chance contingency table.

a) What is the DF value?

b) What is the new χ^2 value?

c) In a general way, what is now the significance of the relationship between visibility and sky blueness?

4. CORRELATION COEFFICIENTS

There are several measures of the closeness of relationships between two variables that represent physical quantities. One is the *rank-difference correlation coefficient*, r_R, which ranges from -1 for a perfect inverse relationship to $+1$ for a perfect direct relationship; a value of zero indicates that there is no relationship. Thus, the larger the absolute magnitude (that is, disregarding the sign) of r_R, the better is the relationship. This is true for all types of correlation coefficients.

As a demonstration of the computation of r_R, we use the first twelve pairs of temperatures and relative humidities in Table III.1. We then

TABLE III.5

Computation of the Rank-Difference Correlation Coefficient for Hourly Temperatures and Relative Humidities in June 1967 at University Park, Pennsylvania

TEMP. (°F)	RH (percent)	TEMP. RANK$_1$	RH RANK$_2$	$R_2 - R_1$	$(R_2 - R_1)^2$
62	55	5	9	4	16
60	61	6	7.5	1.5	2.25
58	69	7.5	6	−1.5	2.25
56	76	9	4	−5	25
54	79	10.5	3	−7.5	56.25
53	83	12	1.5	−10.5	110.25
54	83	10.5	1.5	−9	81
58	72	7.5	5	−2.5	6.25
63	61	4	7.5	3.5	12.25
69	54	3	10	7	49
74	43	2	11	9	81
77	37	1	12	11	121
			Sum	36.0	562.50
				−36.0	
				0.0	

assign the rank of 1 to the highest temperature, the rank of 2 to the next highest temperature, and so on. If there are two or more identical values, we assign to each of these the average of the ranks they would occupy. For example, if three equal values occupy the ranks 7, 8, and 9, we assign the rank of 8 to each and skip the rank numbers 7 and 9, continuing with rank 10 for the next lower value. The same ranking method is applied to the second variable, as shown in Table III.5. We then subtract the ranks R_1 from the respective ranks R_2; for a check: The algebraic sum of the rank differences must be zero. The rank differences are then squared and added.

The correlation coefficient is

$$r_R = 1 - \frac{6\Sigma(R_2 - R_1)^2}{n(n^2 - 1)},$$

where n is the number of pairs; in our example $n = 12$. The sum of the squares of the rank differences, $\Sigma(R_2 - R_1)^2$, is 562 according to Table III.5. Thus,

$$r_R = 1 - \frac{6 \times 562}{12(144 - 1)} = 1 - \frac{562}{286} = 1 - 1.96 = -0.96.$$

This value shows a very high inverse (negative) relationship that was already evident from the small amount of scatter about a straight line in Figure III.2. That is to say, for these data, when the temperature goes up, the relative humidity goes down.

PROBLEM 3

Compute the r_R for the twelve pairs of temperature and relative humidity of 2 May given in Table 6.8. (Use Table III.6.)

a) Compare the result with that obtained from the scatter diagram of Problem 1.

b) Does the magnitude of the correlation coefficient indicate a good or a poor relationship?

In some statistical analyses, the two variates under investigation may have only two categories each, so that only a two-by-two contingency table may be available. In such a case, not only the χ^2 test can be applied to test the significance of the relationship, but the *tetrachoric correlation coefficient*, r_t, can be computed to determine the intimacy of the relationship. As an example of two-category variables, we may consider the frequencies of temperatures above and below 32°F, or of days with

precipitation versus days without precipitation, or hours with relative humidities above 50 percent versus those below 50 percent, and so on. Simultaneous pairs of such bivalued variates are then tallied into a contingency table of the same type as that in Problem 2 (Table III.7).

TABLE III.7

Two-by-Two Contingency Table for Visibility and Sky Blueness, for Computation of the Tetrachoric Correlation Coefficient

B \ V	1 + 2	3 + 4
1 + 2	*A* 61	*B* 14
3	*C* 9	*D* 46

The tetrachoric correlation coefficient is computed with either of these formulas:

$$r_t = \sin\left[90° \frac{\sqrt{A \cdot D} - \sqrt{B \cdot C}}{\sqrt{A \cdot D} + \sqrt{B \cdot C}}\right]$$

or

$$r'_t = \cos\left[180° \frac{\sqrt{B \cdot C}}{\sqrt{A \cdot D} + \sqrt{B \cdot C}}\right]$$

Substituting the frequencies in Table III.7, we obtain, for example,

$$r_t = \sin\left[90° \frac{\sqrt{61 \cdot 46} - \sqrt{14 \cdot 9}}{\sqrt{61 \cdot 46} + \sqrt{14 \cdot 9}}\right] = \sin\left[90° \frac{\sqrt{2806} - \sqrt{126}}{\sqrt{2806} + \sqrt{126}}\right]$$

$$= \sin\left[90° \frac{53.0 - 11.2}{53.0 + 11.2}\right] = \sin\left[90° \frac{41.8}{64.2}\right] = \sin 58.6° = \underline{0.85}.$$

This correlation coefficient is very good and confirms the previously established high significance of the relationship between visibility and blueness of the sky.

TABLE III.6
Computation of r_R for Problem 3 Using Bihourly Temperatures and Relative Humidities of 2 May 1967 at University Park, Pennsylvania

TEMP.	RH	R_1	R_2	$R_2 - R_1$	$(R_2 - R_1)^2$
67	80				
67	90				
63	93				
60	94				
63	93				
65	82				
72	65				
70	100				
64	100				
53	96				
52	87				
51	88				

appendix IV
review

In this last section, a review of many of the preceding topics is presented in the form of questions and problems that refer to the material given in Table IV.1 of comparative climatic data of twenty stations in the conterminous United States. In the questions, the numbers in parentheses refer to the column numbers given at the bottom of the several portions of the table.

1) Why is the mean annual temperature (6) greater at Houston than at St. Louis?
2) Why is the daily range of temperature in July (4,5) larger at Minneapolis than at Boston?
3) Why is the daily temperature range in Minneapolis greater in July (4,5) than in January (2,3)?
4) Compare the mean annual temperature (6) at Denver with that at Omaha; why is Denver, which is farther south than Omaha, colder?
5) Oklahoma City averages 72 days per year with temperatures of 90°F or above (37), while New Orleans, farther south, averages only 57 such days. Why?
6) Which two stations have the largest amounts of precipitation in the extreme wettest month (12)? Why?
7) Can you explain why Miami is one of the few stations that have not recorded a temperature of 100°F or more (7)?
8) Why does Buffalo have a greater mean annual snowfall (16) than any other station listed?
9) Which station has the greatest difference between the extreme highest (7) and lowest (8) temperatures? Which one has the smallest difference? Give the reasons for this.
10) Which are the three windiest stations (24,25)? Which are the three least windy stations?
11) Why can the general increase of wind speed with height shown in Section 7.2 not be verified from the data (1,24,25)?

12) Explain the difference in wind climate (24–26) at Miami and Los Angeles.
13) Arizona has a reputation for much sunshine (27–29). Do the data support that reputation? What are the reasons for this climatic feature?
14) Plot the annual percent sunshine (29) versus the number of clear days (30) on a scatter diagram (see Appendix III). Do the same for the number of cloudy days (32). What can be said about these two relationships if any? Compute for each the rank-difference correlation coefficient (see Appendix III); is the magnitude of the coefficient larger for sunshine/clear days or for sunshine/cloudy days? Explain.
15) What is the range of record-highest temperatures (7) among the various stations? What is the range of record-lowest temperatures (8) among the stations? Explain the large difference between the two ranges.
16) Why do most stations have a lower relative humidity during midday (19,22) than in the morning and evening (18,20 and 21,23)?
17) Compute the average daily relative humidity for each station in January by doubling the 1:30 p.m. value (19), adding the other two values (18,20) and dividing by 4; this gives a weighted average. Do the same for July. The mean daily relative humidities are less in July than in January at all stations except two; identify these stations and explain the reason for this reversal.
18) At what station is the range of mean daily relative humidity between January and July greatest? At what station least? Which two stations have the highest overall average daily relative humidity? Which two stations have the lowest?
19) New York, Pittsburgh, and Omaha lie at approximately the same latitude. Why, then, is there such a pronounced difference in the percent sunshine (27–29) and corresponding cloudiness (30–32)?
20) How can you explain the fact that in spite of having the largest number of days with temperatures of 90°F or above (37), Phoenix does not have the highest average annual temperature (6)?
21) At which station does snow (16) represent the largest percentage of the total precipitation (11)? Explain the situation there.
22) At which station is the difference between the wettest (9) and driest (10) months greatest? At which station is it smallest? What does this mean in terms of the general precipitation climate?
23) Why is there such a large spread of extreme wettest months (12) among stations, whereas the spread is very small for the extreme driest months (13)?
24) Is there a relationship between the extreme wettest months (12) and

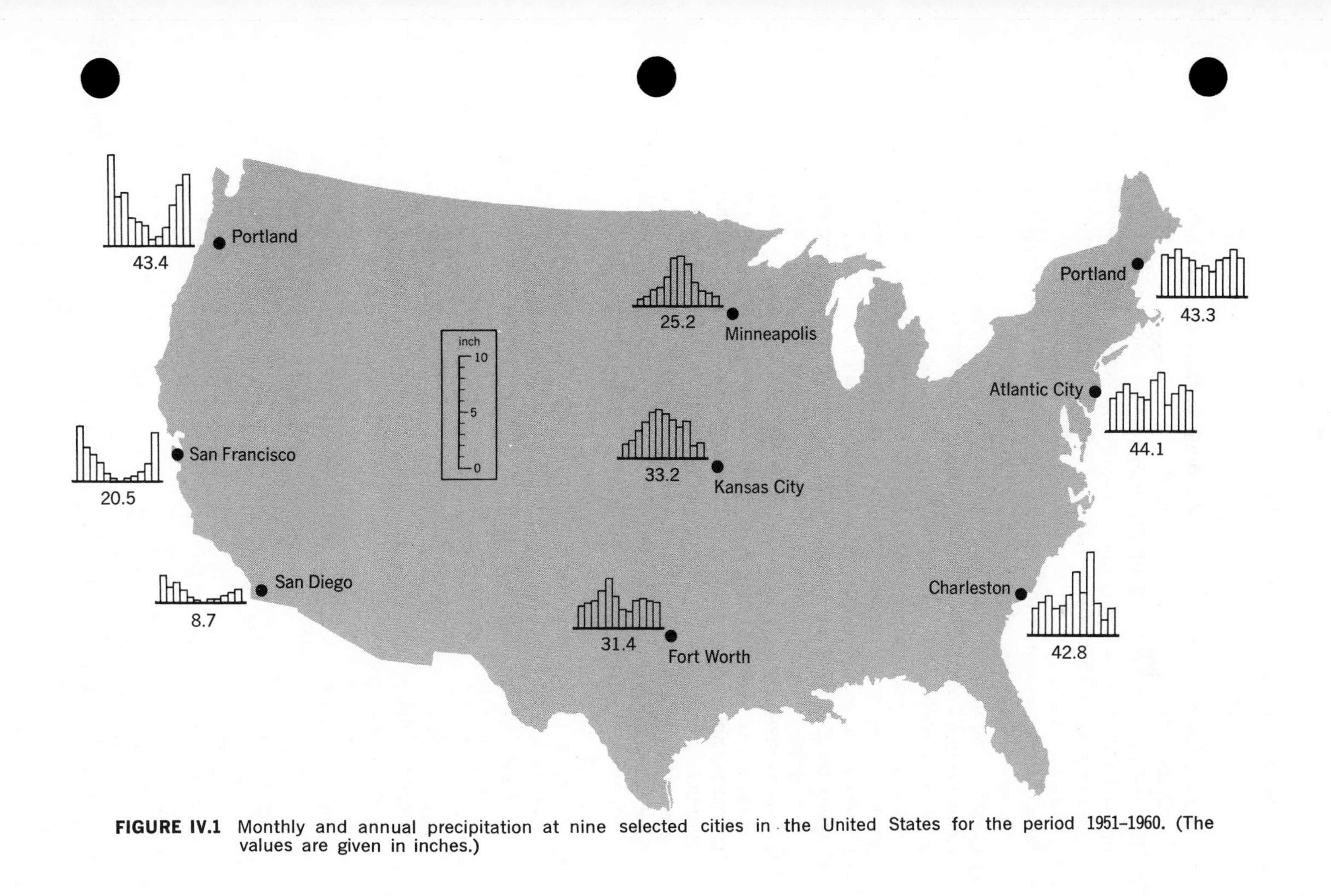

FIGURE IV.1 Monthly and annual precipitation at nine selected cities in the United States for the period 1951–1960. (The values are given in inches.)

the maximum 24-hour rainfall (14)? What is the rank-correlation coefficient? Can you explain the result?

25) There is a very close relationship between the January (15) and seasonal (16) snowfall; why is the relationship between the 24-hour snowfall (17) and both the January (15) and seasonal (16) snowfall only very modest?
26) Why can one not expect the fastest wind speed (26) to be correlated with the mean hourly wind speeds (24,25)?
27) Compare the mean annual temperature ranges (2,3 and 4,5) at the following pairs of stations: Los Angeles and Seattle; Oklahoma City and Minneapolis; Atlanta and Buffalo. What do the temperature ranges at these station pairs have in common? On the other hand, what differences can be noted and what are the reasons for them?
28) In Figure IV.1 each triplet of stations from west to east lies on approximately the same latitude. The bar diagrams represent the mean monthly precipitation starting with January; the scale is shown in the insert. The numbers at each of the stations represent the mean annual totals of precipitation. Interpret and explain the several precipitation climates and their differences from one another in terms of: a) availability of precipitable water; b) seasonal factors such as jet-stream migration, hurricane and thunderstorm occurrences, and so on; c) terrain effects.

TABLE IV.1
Comparative Climatic Data

Station & State	Elevation (feet)	Temperature (°F) January Daily Max.	January Daily Min.	July Daily Max.	July Daily Min.	Annual	Record Highest	Record Lowest
1. Atlanta, Ga.	975	53.3	35.9	89.4	69.6	62.2	103	−9
2. Boston, Mass.	15	36.5	21.6	80.0	64.3	50.7	104	−18
3. Buffalo, N.Y.	693	32.4	18.5	81.1	60.0	47.5	99	−21
4. Chicago, Ill.	610	32.7	17.1	85.3	63.9	50.1	105	−23
5. Cincinnati, Ohio	761	40.9	25.2	87.2	65.9	54.9	109	−17
6. Denver, Colo.	5292	41.7	15.6	87.2	58.3	49.8	105	−29
7. Havre, Mont.	2488	26.3	6.1	85.9	56.6	43.6	108	−57
8. Houston, Tex.	41	61.9	45.7	92.3	75.2	70.0	108	5
9. Los Angeles, Calif.	312	64.6	45.4	83.2	61.7	63.9	110	28
10. Miami, Fla.	8	74.4	62.6	86.8	76.4	75.3	95	27
11. Minneapolis, Minn.	830	23.1	6.1	84.9	63.3	45.6	108	−34
12. New Orleans, La.	9	63.5	48.3	90.3	75.8	70.4	102	7
13. New York, N.Y.	10	39.9	25.8	82.3	66.9	53.4	102	−14
14. Oklahoma City, Okla.	1280	47.0	27.2	93.3	70.9	60.4	109	−10
15. Omaha, Neb.	978	32.4	13.5	89.3	67.6	51.6	114	−32
16. Phoenix, Ariz.	1114	64.8	34.5	105.1	75.1	69.4	118	16
17. Pittsburg, Pa.	1151	36.6	21.3	82.8	61.8	50.6	103	−20
18. St. Louis, Mo.	465	41.0	25.6	89.6	71.6	57.3	112	−22
19. Seattle, Wash.	14	45.2	36.2	75.1	56.1	53.2	100	3
20. Washington, D.C.	72	44.0	28.9	87.1	68.4	56.8	106	−15
Column Number	(1)	(2)	(3)	(4)	(5)	(6)	(7)	(8)

Comparative Climatic Data (Cont'd)

	Precipitation (inches)								
	Normals			Extremes			Snow, Sleet, Hail		
Station Number	Wettest Month	Driest Month	Annual Total	Extreme Wettest Month	Extreme Driest Month	Maximum in 24 hr	January Mean	Seasonal Mean	Maximum in 24 hr
1.	5.67	2.60	49.16	15.82	0.02	7.36	0.9	2.2	8.3
2.	3.50	2.79	38.76	17.09	T*	8.40	11.8	41.8	16.5
3.	3.09	2.43	32.29	10.63	0.05	4.23	18.9	74.8	24.3
4.	4.15	1.41	32.72	12.06	0.06	6.19	8.6	33.8	14.9
5.	4.07	2.19	39.34	13.68	0.17	4.77	4.8	17.8	11.0
6.	2.20	0.50	14.20	8.57	0.00	6.53	6.0	56.1	23.0
7.	2.98	0.39	12.31	9.67	T	3.71	7.0	36.8	24.8
8.	4.84	2.81	45.37	17.64	T	10.83	0.1	0.2	3.0
9.	3.37	T*	14.54	15.80	0.00	7.36	T*	T*	2.0
10.	7.88	1.73	47.20	25.34	T	15.10	0.0	T	T*
11.	4.26	0.80	24.71	11.87	T	7.80	8.9	42.4	16.2
12.	7.09	3.66	63.54	25.11	T	14.01	0.1	0.2	8.2
13.	4.34	3.04	42.03	14.51	0.11	9.40	7.5	30.1	25.8
14.	4.25	1.24	30.22	10.78	T	4.82	3.8	9.1	8.4
15.	4.51	0.81	25.90	12.70	T	7.03	6.5	27.9	16.4
16.	1.00	0.06	7.16	6.47	0.00	4.98	T	T	1.0
17.	4.07	2.37	36.92	10.25	0.06	4.08	8.5	33.0	17.5
18.	4.10	1.88	37.86	20.45	0.00	8.78	4.2	17.1	20.4
19.	5.34	0.52	31.92	15.33	0.00	3.52	5.0	11.2	21.5
20.	4.49	2.64	41.44	17.45	0.14	7.31	6.0	19.5	25.0
Col. No.	(9)	(10)	(11)	(12)	(13)	(14)	(15)	(16)	(17)

T* = Trace, but not measurable amount.

Comparative Climatic Data (Cont'd)

STATION NUMBER	RELATIVE HUMIDITY (percent)						WIND SPEED, MEAN HOURLY		mph
	JANUARY			JULY					
	7:30 a.m.	1:30 p.m.	7:30 p.m.	7:30 a.m.	1:30 p.m.	7:30 p.m.	JAN.	JULY	FASTEST MILE
1.	80	64	69	83	57	68	11.5	7.8	70
2.	71	59	67	72	55	70	12.8	10.1	87
3.	79	72	78	78	53	63	17.5	12.1	91
4.	80	70	74	76	55	61	11.8	8.7	87
5.	82	70	74	84	52	60	8.6	5.3	49
6.	58	42	47	63	30	33	7.7	7.1	65
7.	82	71	77	75	38	35	9.4	7.5	59
8.	85	66	73	90	58	66	10.6	8.4	84
9.	65	45	57	87	52	58	6.4	5.8	49
10.	87	59	75	83	64	77	13.5	10.4	132
11.	77	67	72	79	51	54	11.2	9.7	92
12.	85	67	73	84	64	72	8.3	6.4	98
13.	72	61	66	75	57	68	16.4	12.0	113
14.	79	63	63	82	52	49	16.1	12.1	87
15.	80	66	69	76	51	52	9.8	8.1	109
16.	70	40	38	53	30	23	4.3	5.8	65
17.	77	67	72	77	52	60	11.6	8.5	73
18.	77	65	68	73	50	55	11.9	9.2	91
19.	86	81	74	86	63	48	10.2	7.9	65
20.	73	56	64	78	52	68	7.7	5.7	62
Col. No.	(18)	(19)	(20)	(21)	(22)	(23)	(24)	(25)	(26)

Comparative Climatic Data (Cont'd)

Station Number	Percent of Possible Sunshine			Annual Mean Number of Days									
				Sunrise to Sunset							Temperature		
											Max.	Minimum	
	Jan.	July	Annual	Clear	Partly Cloudy	Cloudy	Precip. 0.01 inch or more	Snow 1.0 inch or more	Thunder-storms	Dense Fog	90° and above	32° and below	0° and below
1.	48	62	62	126	111	128	122	0	50	21	33	38	0
2.	49	64	57	114	115	136	126	11	19	15	9	103	2
3.	31	70	50	72	130	163	164	22	29	13	2	128	3
4.	44	73	58	113	120	132	124	10	37	11	13	110	8
5.	41	72	57	108	115	142	135	7	52	17	30	95	1
6.	67	69	67	142	152	71	86	17	43	4	23	139	8
7.	49	78	62	125	132	108	90	13	23	5	19	171	40
8.	49	69	60	116	126	123	101	0	57	16	83	8	0
9.	70	78	72	181	124	60	39	0	5	23	14	0	0
10.	66	65	66	103	155	107	133	0	72	2	6	0	0
11.	49	72	56	105	113	147	107	15	38	8	15	149	31
12.	49	58	59	119	137	109	119	0	74	14	57	4	0
13.	51	65	60	105	130	130	125	8	31	18	7	91	0
14.	59	79	67	152	97	116	83	2	51	15	72	76	0
15.	56	77	62	133	117	115	98	10	40	9	31	123	14
16.	76	83	84	228	89	48	36	0	26	1	168	21	0
17.	32	65	50	85	128	152	150	11	40	24	16	103	2
18.	48	72	59	138	115	112	112	5	49	10	37	80	2
19.	25	64	43	74	106	185	152	3	5	24	1	20	0
20.	46	64	57	125	119	121	124	6	33	11	27	81	0
Col. No.	(27)	(28)	(29)	(30)	(31)	(32)	(33)	(34)	(35)	(36)	(37)	(38)	(39)

authors' note

In keeping with current practices in the United States, both the English and the metric systems have been used as units of length, speed, temperature, and so on. No attempt has been made at converting the data to the same system, but the units have usually been left in the form in which they were collected or presented in the original sources. Inasmuch as no universal standard has been adopted as yet in this country, we believe that the dual use of units will induce the student to become proficient in converting one unit into the other. The following brief table should facilitate these conversions.

METRIC UNITS	ENGLISH UNITS
1 km = 1000 m = 0.621 stat. mile	1 stat. mile = 0.868 naut. mile
= 0.540 naut. mile	= 5280 ft = 1.61 km
1 m = 100 cm = 3.28 ft	1 naut. mile = 1.15 stat. miles
1 cm = 10 mm = 0.394 inch	= 6080 ft = 1.85 km
1 mps = 2.24 mph = 1.94 knots	1 ft = 30.5 cm
	1 inch = 2.54 cm
1 deg. latitude = 60.0 naut. miles	1 mph = 0.868 knots = 0.477 mps
= 69.1 stat. miles	= 1.61 kmph
= 111 km	1 knot = 1 naut. mph = 1.15 mph
	= 0.515 mps = 1.85 kmph

Units may be converted from one system to the other by multiplication with the appropriate factor given above. Other conversions may be made by multiplying by fractions that are equal to unity; for example, in the case of conversion of solar radiation units from 247,000 langleys per year into units of langleys per day, one multiplies by (1 year/365 days = 1) so that

$$\frac{247{,}000 \text{ langleys}}{\text{year}} \times \frac{1 \text{ year}}{365 \text{ days}} = 677 \text{ langleys/day.}$$

index